SpringerBriefs in Animal Sciences

More information about this series at http://www.springer.com/series/10153

Roman Fuchs • Petr Veselý • Jana Nácarová

Predator Recognition in Birds

The Use of Key Features

 Springer

Roman Fuchs
Department of Zoology, Faculty of Science
University of South Bohemia
České Budějovice, Czech Republic

Petr Veselý
Department of Zoology, Faculty of Science
University of South Bohemia
České Budějovice, Czech Republic

Jana Nácarová
Department of Zoology, Faculty of Science
University of South Bohemia
České Budějovice, Czech Republic

ISSN 2211-7504   ISSN 2211-7512   (electronic)
SpringerBriefs in Animal Sciences
ISBN 978-3-030-12402-1   ISBN 978-3-030-12404-5   (eBook)
https://doi.org/10.1007/978-3-030-12404-5

Library of Congress Control Number: 2019934444

This Springer imprint is published by the registered company Springer Nature Switzerland AG.
The registered company address is: Gewerbestrasse 11, 6330 Cham, Switzerland

Abstract

In the first chapter, we summarize methodological approaches in the research of predator recognition. In the second chapter, we summarize results of studies showing the ability of birds to differentiate predators from harmless animals, particular predators from different ecological guilds (predators of adults and of chicks) and even particular predator species. In the third chapter, we describe the means used by birds during the recognition process. Most of the studies prove the importance of so-called key features. Some studies altering multiple features suggest that the perception of predators is rather complex in birds. In the fourth chapter, we try to link the knowledge on predator recognition by untrained birds and the psychological point of view based on studies observing the recognition process of birds trained with the use of operant conditioning. Given together, such studies show that birds are surprisingly flexible during the recognition process and are able to respond according to the features available for recognition. For future research, we find great potential in studies focusing on assessing the relative importance of different feature types and processes used by birds in object categorization.

Keywords Antipredatory behaviour · Raptor · Owl · Passerine · Nest defence · Mobbing · Categorization · Learning · Cognition

Acknowledgements

We wish to thank all the students of our research group, who participated in the research of predator recognition and provided the foundation for our knowledge on this topic. We especially thank Michal Němec and David Nácar for their help with the first drafts of the text. We are grateful to Michaela Syrová for checking the spelling and syntax of the manuscript.

Contents

Chapter 1
How to Study Predator Recognition

If we wish to learn how preys recognize their predators, we have to accomplish two tasks. In the first place, their mutual encounter must be arranged. We can either search for spontaneous predatory events in the wild, which is rather inefficient, requiring expenditure of quite an effort from the observer (particularly in terms of the required amount of time), or we can employ an experimental approach. The meeting of a prey and a predator can be organized both in the wild and in captivity. Both methods have their strengths and weaknesses. What brings problems in the wild is the standardization of experiments, as the experimental animals are never under our full control. In captivity, the representativeness of experiments is mainly a problem, since the conditions here never fully tally with those in the wild.

If we have arranged an encounter of the prey and the predator, we have yet to find sensitive and reliable markers indicating that the prey has recognized the predator. A sensitive marker should detect not only whether the prey distinguishes the predator from a harmless animal (or even from another neutral object), but also various groups or even kinds of predators from each other. A reliable marker should yield identical results at each meeting of the predator and the prey. The second requirement in particular is undoubtedly a maximalist one. The reaction of the prey is influenced by its own situation (e.g. physical condition) on the one hand and by external factors (e.g. the availability of shelters) on the other. Taking into account these complications, physiological parameters indicative of fear seem to be suitable markers, since undeniably they are more directly linked with the process of recognizing the predator than a behavioural response. Their use, however, is complicated by methodical limitations. Therefore, directly observable behavioural markers, using a variety of defensive reactions, play and will keep playing a major role.

Researching how birds recognize their predators has never been very intense; nonetheless, it has a long tradition dating back to the very beginning of behavioural ecology and ethology. On that account, the existing studies differ not only in their methodologies but also in the correctness of the design. This chapter aims to provide an overview of the procedures applied in researching the recognition of predators,

© The Author(s) 2019
R. Fuchs et al., *Predator Recognition in Birds*, SpringerBriefs in Animal Sciences,
https://doi.org/10.1007/978-3-030-12404-5_1

compare their strengths and weaknesses and try to assess their impact on the results achieved.

1.1 Observation in the Wild

The majority of animals are under the risk of predation for most of their lives; therefore, antipredator behaviour is an integral part of their everyday life (Caro 2005). However, antipredator behaviour involves a number of partial consecutive elements (Lima and Dill 1990). Primarily, animals on the edge of the flock or herd or solitary individuals spend a considerable part of the day being vigilant, monitoring their surroundings in order to detect the predator before it discovers them (Elgar 1989). If successful, they have a chance to avoid the predator without directly confronting it (Sansom et al. 2009). Subsequently, the observer has no chance of noticing any interaction between the predator and the prey.

Providing the prey does encounter the predator, it is usually very short (e.g. Smith 1970), and therefore the observer is unlikely to be present at that particular moment. In addition to that, escape is a universal response in case of birds (Simmons 1955; Lima 1993; Hilton et al. 1999; Martin et al. 2006; van den Hout et al. 2010a). The most common interaction of birds with predators thus usually provides little information about their ability to recognize predators.

However, birds are very distinctive with one manifestation of their behaviour, which enables us to assess by simply observing to what extent they recognize predators, and that is the nest defence. A nest with eggs—or even more a nest with chicks—presents an extremely valuable and at the same time vulnerable object for the parents. Naturally, the chicks in the nest have only a limited possibility to defend themselves (Redondo and Carranza 1989; Goławski and Meissner 2007; Hagelin and Jones 2007; Tillmann 2009; Londoño et al. 2015). Therefore, parental investment into defending the nest significantly increases their fitness (Knight and Temple 1986a; Redondo 1989; Tryjanowski and Goławski 2004; Müller et al. 2005; Remeš 2005; Goławski and Mitrus 2008). This is particularly true for birds of temperate and cold climate zones, which usually have only a limited opportunity of substitute nesting (Skutch 1949) and whose survival up to the next season is often uncertain whether they migrate or are residents. Nevertheless, even in the case of defending the nest, the main form of its defence is not to draw the predator's attention to it. This can be primarily achieved by being cautious when building the nest as well as during the incubation of eggs and feeding the young ones (Dale et al. 1996; Burhans 2000; Ghalambor and Martin 2000; Roos and Pärt 2004; Eggers et al. 2005; Amo et al. 2008; Peluc et al. 2008). However, if the nest is discovered by the predator, active defence called 'mobbing' usually remains the only way to prevent predation (Montgomerie and Weatherhead 1988). The active defence usually takes longer than a flight, also including a larger number of activities varying in risk, which the defending birds carry out (Sordahl 1990). Choosing which one to use depends, besides other aspects, on the dangers that the predator present poses for the defending parents and/or for their offspring (Ash 1970; Curio 1975; Gottfried 1979; Curio et al. 1983; Curio and Regelmann 1985; Elliot 1985; Curio and Onnebrink 1995; Hogstad 2005).

With most birds, predation represents the most common cause of nesting failures (Ricklefs 1969; Nilsson 1984; Martin and Roper 1988; Martin 1993a, b), the proportion of predated nests ranging from 1.4% (Holway 1991) in the case of well-concealed nests of the Black-throated Blue Warbler (*Setophaga caerulescens*) to 85% in the case of the Northern Cardinal (*Cardinalis cardinalis*—Filiater et al. 1994) or some South American species of songbirds (Skutch 1996). Still, the probability of capturing a predation attempt on a particular nest remains rather low as such an occurrence usually lasts no more than minutes (Weidinger 2010). In recent decades, the possibility of continuous nest monitoring by using video cameras has arisen (reviewed by Cutler and Swann 1999 and more recently by Cox et al. 2012); nonetheless, it has not been used for studying the defending birds' recognition of predators heretofore, one possible cause being that only few specialized predators cause the nesting losses with most birds (e.g. Conner et al. 2010; Conkling et al. 2012; Friesen et al. 2013; Murray 2015).

In spite of that, monitoring the encounters of nesting birds and predators brought some pieces of knowledge regarding their recognition, albeit it may not have been the main objective of the studies in question. Very often, these were ground-nesting species, whose nests in an open terrain can be, at least theoretically, easily discovered by all passing birds. These characteristics are often met by members of the Charadriiformes order, out of which the following have been studied: the plovers (Brunton 1986, 1990; Byrkjedal 1987; Amat and Masero 2004), the lapwings (Green et al. 1990; Walters 1990), the godwits (Green et al. 1990), the stilts and avocets (Sordahl 2004) and the gulls and terns (Kruuk 1964; McNicholl 1973; Cavanagh and Griffin 1993; Brunton 1997, 1999; Palestis 2005; Stenhouse et al. 2005), but also, e.g. the cormorants (Siegel-Causey and Hunt 1981) and ducks (Jacobsen and Ugelvik 1992).

The effectiveness of such researches was strengthened in most of the studied species by the fact that their breeding is colonial or semicolonial. Similar studies could therefore be made on colonially breeding songbirds as well, e.g. the swallows (Guillory and LeBlanc 1975; Winkler 1992) or momots (Murphy 2006). A breeding colony is undoubtedly more noticeable for predators (and researchers for that matter) than an individual nest, representing also a more attractive prey, with another advantage being that it allows you to monitor the behaviour of a large number of individuals differing in the degree of imminent threat as well as in the motivation to defend, which primarily depends on the value of their offspring as far as the age and the number of chicks or eggs are concerned (Barash 1976; Andersson et al. 1980; Knight and Temple 1986b; Conover 1987; Montgomerie and Weatherhead 1988; Burger et al. 1993; Clode et al. 2000). The number of birds involved in the antipredator behaviour can, for instance, be employed as a criterion of the extent to which a particular predator was considered a danger (Fuchs 1977; Shields 1984; Brown and Hoogland 1986; Burger and Gochfeld 1992; Clode et al. 2000; Arnold 2000; Bosque and Molina 2002).

However, besides these colonially breeding species, the ability to recognize naturally occurring predators even when defending the nests was observed with solitarily breeding species, such as the sparrows (Nice and Ter Pelkwyk 1941), magpies (Buitron 1983) and phainopeplas (Leger and Carroll 1981).

To observe natural antipredator behaviour, it is not needful to watch nesting birds. Trail (1987), for instance, observed the behaviour of Guianan Cock-of-the-rock (*Rupicola rupicola*) towards various potential predators on leks, where the males collectively utter the mating call. Foraging constitutes another situation in which birds are readily observable, one advantage being that some groups of birds group together in order to search for food. In the temperate zone of the northern hemisphere, e.g. winter, flocks of songbirds have been observed, especially the tits (Hill 1986; Gentle and Gosler 2001; Davies and Welbergen 2008; Soard and Ritchison 2009; Courter and Ritchison 2010; Tvardíková and Fuchs 2010, 2011, 2012; Suzuki 2011), but also the fringillid (Whittingham et al. 2004; Quinn and Cresswell 2005) and the corvid birds (Hauser and Caffrey 1994; Griesser 2008, 2009). Similarly, you can also use foraging flocks of the starlings (Conover and Perito 1981), pigeons (Griffin et al. 2005) or waders (Minderman et al. 2006; Mathot et al. 2009). Regrettably, under normal circumstances, such flocks do not provide the opportunity of continuous monitoring, since they are very agile not only over long time but also during the day (Tvardíková and Fuchs 2010, 2011, 2012).

What brought interesting data regarding the ability to recognize predators were studies where authors systematically observed species noticeable due to their behaviour, such as the drongos (*Dicrurus*), which hunt using the sit-and-wait strategy, thus being very well visible in the undergrowth of a tropical forest (Nijman 2004). Despite that the study was still based on 8 years of data collection. When studying the socially living Arabian babblers (*Turdoides squamiceps*—Edelaar and Wright 2006), the authors used the fact that a migration of Palaearctic predators runs through this location in their research in the Israeli Arava Rift Valley. During the 9-week research (1–6 h a day), they managed to gather almost 250 contacts of babbler flocks with potential predators.

1.2   Experiments

It is evident that the observational approach only provides very limited possibilities for studying predator recognition. Most of the existing pieces of knowledge therefore come from different types of experiments, which primarily increase the efficiency of research by staging a meeting of a prey and a predator, the latter not being an accidently present one, but an intentionally selected predator. If the research focuses on the recognition process, an experimental approach is of an essential importance, which requires manipulation of potential recognition features enabled by decoy predators (Edwards et al. 1950; Curio 1975; Scaife 1976; Smith and Graves 1978; Gill et al. 1997a; Davies and Welbergen 2008; Trnka et al. 2012; Beránková et al. 2014).

Nonetheless, avoiding encounters with predators as the preferred form of antipredator behaviour applies even to experiments (Krätzig 1940; Scaife 1976; Palleroni et al. 2005). Admittedly, it is generally easy—at least in natural experiments, being all the more true for birds characterized by exceptionally high mobility. An accidental exposure of a predator in the field can thus hardly bring any usable results. There is one exception, though. Birds usually actively react to the presence of the owls during the day, regardless of any other circumstances (Naguib et al. 1999; Yorzinski and Vehrencamp 2009), since the owls constitute major enemies for them, being temporarily handicapped during the day, though. An active defence is aimed to drive them away from the territory or home area in order to prevent any night-time peril. This method was repeatedly used for studying predator recognition both in studies comparing the response to the owls and other predators (Naguib et al. 1999; Yorzinski and Vehrencamp 2009) and in studies comparing different owl species (Miller 1952; Altmann 1956; Reudink et al. 2007). This method was even used to evaluate a danger represented by a modified predator. Deppe et al. (2003) presented wooden decoys of the Northern pygmy owl (*Glaucidium gnoma*) on clearings in the forests near the town of Missoula (Montana, USA). The decoys varied in the presence of eyespots on the nape of the pygmy owl. The authors monitored which sides were the mobbing birds living on that particular site coming from. Nonetheless, the testing of the same species, let alone individuals, cannot be guaranteed in individual trials, which represents the principal disadvantage of this approach. There are a few studies that even presented birds of prey (Rainey et al. 2004), the shrikes (Chu 2001), or even mammalian predators (Naguib et al. 1999; Rainey et al. 2004; Randler 2006) to birds in this experimental arrangement. Sometimes these works also focused on a particular species under study (Naguib et al. 1999; Rainey et al. 2004; Adams et al. 2006; Randler 2006; Magrath et al. 2007).

However, the advantage of presenting predators in the wild is that we can manage to make the tested animals not to opt for the preferred form of antipredator behaviour, i.e. avoiding predators. This can be achieved by inducing a situation in which a trade-off between avoiding a predator and other interest arises (Caro 2005).

1.2.1 Experiments on Nests

As already mentioned in Sect. 1.1, nest defence is perhaps the most important situation for birds. However, even in defending the nest, the effort to prevent its discovery prevails. For example, the red-backed shrikes (*Lanius collurio*) significantly reduce the intensity of active defence as soon as a mounted jay is located at a distance of 10 m from the nest (Tichá unpublished observation). In field experiments, the predator is therefore usually located in the immediate vicinity of the nest, and in the case of decoys, the gaze is directed towards the nest in order to induce an impression of an acute threat. The experiments are usually carried out at the chick feeding stage, when parents respond most actively. This is due to the fact that for many reasons, the value of the offspring increases during the nesting season (Knight and Temple 1986c; Montgomerie and Weatherhead 1988; Amat et al. 1996; Halupka 1999; Campobello and Sealy 2010), with the difficult concealment of the nest during the feeding stage being also of a considerable importance (Rytkönen et al. 1995; Goławski and Mitrus 2008).

Despite all the efforts to standardize, the response of the tested birds usually varies, the cause being various parameters of the parents, in addition to a different value of the offspring determined by the number and age of the chicks (Andersson et al. 1980; Montgomerie and Weatherhead 1988; Halupka and Halupka 1997; Halupka 1999; Pavel and Bureš 2001; Tryjanowski and Goławski 2004), as well as by their sex ratio (Radford and Blakey 2000) and their chances for being successfully raised and surviving (Hakkarainen and Korpimäki 1994). Above all, the following influences were ascertained: the influence of a different age or more precisely experience (Montgomerie and Weatherhead 1988), of a sex (Wiklund 1990; Tryjanowski and Goławski 2004; Hogstad 2005), of a current physical condition (Hamer and Furness 1993; Hogstad 1993, 2005; Griggio et al. 2003), of a hormonal state (Cawthorn et al. 1998) and also of the personality of the parents (Hollander et al. 2008). Nonetheless, the intensity of the defence can also be affected by the environment around the nest (Kleindorfer et al. 2005) and its concealment and accessibility (Montgomerie and Weatherhead 1988; Holway 1991; Weidinger 2002; Kleindorfer et al. 2003; Remeš 2005; Goławski and Mitrus 2008). Pairs that fail to react to a clearly dangerous predator or only react very weakly are quite a common occurrence as well (Strnad et al. 2012). Besides the aforementioned factors, even the unreliability of the experiment cannot be ruled out as a cause, especially in the case of using decoy predators (Knight and Temple 1986b; Weatherhead 1989; Rytkönen et al. 1990; Grim 2005; Němec et al. 2015). To obtain credible results therefore requires a relatively large amount of data, i.e. the number of pairs tested, with the number in the present-day works ranging from 13 (Csermely et al. 2006) to 120 pairs (Arroyo et al. 2001).

Despite certain drawbacks, experiments simulating nest threats are widely used, not only in researching predator recognition. The ability of birds to optimally solve the trade-off between the investment in defence and the offspring's value for the parents was perhaps studied most intensively in this way (Greig-Smith 1980; East 1981; Burger et al. 1989; Rytkönen et al. 1990, 1995; Hamer and Furness 1993; Curio and Onnebrink 1995; Ghalambor and Martin 2000; Rytkönen 2002; Fischer and Wiebe 2006). Researching the predator recognition ability is an integral part—albeit often not the main addressed issue—of another fairly large group of works aimed at the trade-off between the investment in defence and the danger a predator poses to the parents and chicks (Tables 1.1 and 1.2). Studies focused on the process of predator (and brood parasite) recognition, particularly on the features the birds are guided by in the process of recognition, constitute the last and smallest group of works using nest experiments (Table 1.3).

Most of the studies dealing with predator recognition and based on nest experiments use songbirds (Tables 1.1–1.3), this probably being mainly due to their higher nest densities, which facilitate obtaining sufficient material. For the same reason, the most popular species among songbirds are the colonially breeding ones (Barash 1976; Smith and Graves 1978; Shields 1984; Brown and Hoogland 1986; Knight

Table 1.1 Summary of studies testing the ability of birds to recognize various predators with the help of changing the features of the stimuli

References	Method	Measured marker	Tested bird	Bird origin	Presented predator	Predator form	Control
Krätzig (1940)	In the lab	Freezing, hiding	*Lagopus lagopus*—chicks	Captive bred	Eagle, falcon, goose flying backwards	Moving silhouette	Geometric shapes, goose
Nice and Ter Pelkwyk (1941)	In the lab	Alarm calls	*Melospiza melodia*	Captive bred	Owl—with and without eyes, head alone	Drawing on a cardboard model	Shaded and empty shape
Scaife (1976)	In the lab	Distance	*Gallus gallus f. domestica*	Captive bred	*Falco tinnunculus*—with/without eyes, with prolonged bill	Stuffed dummy	*Apteryx australis*
Klump and Curio (1983)	In the lab	Freezing, alarm calls	*Cyanistes caeruleus*	Nature	*Accipiter nisus*—small and large	Moving silhouette	None
Alatalo and Helle (1990)	In the lab	Alarm calls	*Poecile montanus*	Wild-caught	*Accipiter nisus*—small and large	Moving silhouette	None
Evans et al. (1993b)	In the lab	Alarm calls, nonvocal response	*Gallus gallus f. domestica*	Captive bred	Raptor-shaped images	Video	None
Schleidt et al. (2011)	In the lab	Alarm calls	*Meleagris gallopavo*	Captive bred	Raptor	Plywood silhouette	Circle, stick
Beránková et al. (2014)	In the lab	Mobbing, fleeing, comfort—index	*Parus major*	Wild-caught	*Accipiter nisus*—salient features modifications	Wooden dummy	*Columba livia f. domestica*
Beránková et al. (2015)	In the lab	Mobbing, fleeing, comfort—index	*Parus major*	Wild-caught	*Accipiter nisus*—size and colouration modifications	Plush dummy	Empty control
Nácarová et al. (2018)	In the lab	Mobbing, fleeing, comfort—index	*Parus major*	Wild-caught	*Accipiter nisus*—salient features modifications	Wooden dummy	*Columba livia f. domestica*

(continued)

Table 1.1 (continued)

References	Method	Measured marker	Tested bird	Bird origin	Presented predator	Predator form	Control
Curio (1975)	At the nest	Mobbing	*Ficedula hypoleuca*	Nature	*Lanius collurio, Glaucidium passerinum*—altered colouration, size, position, salient features	Stuffed and plastic dummy	None
Smith and Graves (1978)	At the nest	Mobbing	*Hirundo rustica*	Nature	*Bubo virginianus*—with/without head, beak, eyes	Stuffed dummy	*Aix sponsa*, artificial objects
Davies and Welbergen (2008)	At the feeder	Feeder attendance	*Parus major, Cyanistes caeruleus*	Nature	*Accipiter nisus, Cuculus canorus*—with/without underpart barring	Stuffed dummy	*Streptopelia decaocto, Anas crecca*
Tvardíková and Fuchs (2010)	At the feeder	Feeder attendance	*Parus major, Cyanistes caeruleus, Poecile* sp.	Nature	*Accipiter nisus*—amputated, occluded	Stuffed dummy	*Columba livia f. domestica*
Veselý et al. (2016)	At the feeder	Feeder attendance	*Parus major, Cyanistes caeruleus*	Nature	*Accipiter nisus*—colouration modifications	Plush dummy	*Columba livia f. domestica*
Nováková et al. (2017)	At the feeder	Feeder attendance	*Parus major, Cyanistes caeruleus, Poecile* sp.	Nature	*Accipiter nisus*—spatial modifications	Modelling material	*Columba livia f. domestica*
Welbergen and Davies (2011)	At the nest	Distance, mobbing	*Acrocephalus scirpaceus*	Nature	*Accipiter nisus, Cuculus canorus*—with/without underpart barring	Stuffed dummy	*Streptopelia decaocto*
Trnka et al. (2012)	At the nest	Mobbing	*Acrocephalus arundinaceus*	Nature	*Cuculus canorus*—with/without under-part barring and yellow eye	Stuffed dummy	*Streptopelia decaocto*
Edwards et al. (1950)	At the nest	Mobbing	*Phylloscopus trochilus*	Nature	*Cuculus canorus*	Wooden dummy with stuffed head, wings and tail	None

Gill et al. (1997a)	At the nest	Alarm calls	*Setophaga petechia*	Nature	*Molothrus ater*—normal/elongated bill	Stuffed dummy	None
Deppe et al. (2003)	In the field	Mobbing	Various bird species	Nature	*Glaucidium gnoma*—with/without eyespots	Wooden dummy with playback	None

Table 1.2 Summary of studies testing the bird ability to recognize predators according to the actual condition in which is the same predator presented

References	Method	Measured marker	Tested bird	Bird origin	Predator	Predator form	Control
Carter et al. (2008)	In the lab	Vigilance	Sturnus vulgaris	Wild-caught	Human—watching/averted	Living	None
Mathot et al. (2009)	In the lab	Vigilance	Calidris canutus	Nature	Accipiter nisus—gliding/sitting	Stuffed dummy moving	None
van den Hout et al. (2006, 2010a)	In the lab	Time spent flying, body mass, pectoral muscle mass	Arenaria interpres, Calidris canutus	Nature	Accipiter nisus—gliding/sitting	Stuffed dummy moving	None
Barash (1976)	At the nest	Mobbing	Corvus brachyrhynchos	Nature	Bubo virginianus—with/without a dead crow	Plastic dummy	None
Shalter (1978)	At the nest	Mobbing	Ficedula hypoleuca	Nature	Glaucidium passerinum, G. perlatum—various types of presentation	Stuffed dummy, living	None
Shedd (1982)	At the nest	Mobbing	Turdus migratorius	Nature	Megascops asio—various types of presentation	Stuffed dummy, living, playback	None
Conover (1987)	At the nest	Mobbing	Larus delawarensis	Nature	Human—with/without gull	Living	None
Watve et al. (2002)	At the nest	Nest visits	Merops orientalis	Nature	Human—not/watching	Living	None
Němec et al. (2015)	At the nest	Mobbing	Lanius collurio	Nature	Garrulus glandarius—various types of presentation	Stuffed, plush and silicone dummy	None
Conover and Perito (1981)	At the feeder	Arrival latency, number of birds	Sturnus vulgaris	Nature	Bubo virginianus—with living/dead prey, playback	Plastic dummy	Empty control, tethered starling
Hill (1986)	At the feeder	Visiting the feeder, distance	Parus bicolor	Nature	Falco columbarius—with/without alarm playback	Stuffed dummy	None

Griesser (2008)	At the feeder	Mobbing calls	*Perisoreus infaustus*	Nature	*Accipiter gentilis, A. nisus*—sitting/flying	Stuffed dummy	Empty control, *Cyanocitta cristata*
Bažant (2009)	At the feeder	Feeder attendance	Various passerine species	Nature	*Accipiter nisus, Martes sp.*—not/moving	Stuffed dummy	*Columba livia f. domestica*, plush toy
Carlson et al. (2017a)	At the feeder	Calling, feeding, scanning, wing beating	*Cyanistes caeruleus*	Nature	*Accipiter nisus* moving, calling, + alarm and distress call	Stuffed dummy	None
Hamerstrom (1957)	In the field	Mobbing	Various bird species	Nature	*Buteo jamaicensis*—hungry/well fed	Living	None
Chandler and Rose (1988)	In the field	Mobbing	Various bird species	Nature	*Megascops asio*—with/without playback	Stuffed dummy and playback	None
Chu (2001)	In the field	Mobbing	Various bird species	Nature	*Lanius ludovicianus*—with playback/alarm call imitations	Plastic dummy and playback	None
Hendrichsen et al. (2006)	In the field	Mobbing	Various bird species	Nature	*Strix aluco*—hidden in vegetation/exposed	Stuffed dummy	None
Cavanagh and Griffin (1993)	Field observation	Mobbing	*Sterna hirundo, Leucophaeus atricilla*	Nature	*Larus argentatus, L. marinus*—flying over/landing	Living	None

Table 1.3 Summary of studies testing the bird ability to recognize various predator species from the same ecological guild

References	Method	Measured marker	Tested bird	Bird origin	Predators, parasites	Predator form	Control
Nice and Ter Pelkwyk (1941)	In the lab	Alarm calls	*Melospiza melodia*	Captive bred	*Molothrus ater*, various predators	Stuffed dummy	*Sturnus vulgaris*
Flasskamp (1994)	In the lab	Mobbing	*Turdus merula*	Wild-caught	*Athene noctua, Strix aluco*	Living	None
Palleroni et al. (2005)	In the lab	Alarm calls, hiding, fleeing	*Gallus gallus f. domestica*	Captive bred—seminatural	*Accipiter striatus, A. cooperii, A. gentilis*	Living	None
Templeton et al. (2005)	In the lab	Alarm calls	*Poecile atricapilla*	Captive bred—seminatural	Ferret, cat, 13 owl and raptor species	Living	None
Azevedo and Young (2006a, b)	In the lab	Alert behaviour	*Rhea americana*	Captive bred	*Panthera onca*, dog	Stuffed dummy, living	Chair
Edwards et al. (1950)	At the nest	Mobbing	*Phylloscopus trochilus*	Nature	*Accipiter nisus, Falco vespertinus*	Stuffed dummy	None
Edwards et al. (1950)	At the nest	Mobbing	*Phylloscopus trochilus*	Nature	*Accipiter nisus, Cuculus canorus*	Stuffed dummy	None
Ash (1970)	At the nest	Mobbing	*Lanius collurio*	Nature	*Tyto alba, Strix nebulosa*	Stuffed dummy	None
Ash (1970)	At the nest	Mobbing	*Lanius collurio*	Nature	*Accipiter nisus, Cuculus canorus*	Stuffed dummy	None
Curio et al. (1983)	At the nest	Mobbing	*Parus major*	Nature	*Strix aluco, Glaucidium perlatum*	Living	None
Curio and Regelman (1985)	At the nest	Mobbing	*Parus major*	Nature	*Strix aluco, Glaucidium passerinum*	Living	None

Duckworth (1991)	At the nest	Mobbing	Acrocephalus scirpaceus	Nature	Accipiter nisus, Cuculus canorus, Garrulus glandarius	Stuffed dummy	None
Neudorf and Sealy (1992)	At the nest	Mobbing	Agelaius phoeniceus, Dumetella carolinensis, Icterus galbula, Bombycilla cedrorum	Nature	Molothrus ater, Quiscalus quiscula	Stuffed dummy	Passerella iliaca
Bazin and Sealy (1993)	At the nest	Alarm calls	Tyrannus tyrannus	Nature	Molothrus ater, Quiscalus quiscula	Freeze-dried dummy	Passerella iliaca
Gill and Sealy (1996)	At the nest	Mobbing	Setophaga petechia	Nature	Molothrus ater, Quiscalus quiscula	Stuffed dummy	Passerella iliaca
Clode et al. (2000)	At the nest	Mobbing	Sterna hirundo, S. paradisaea, Larus fuscus, L. argentatus, L. canus	Nature	Neovison vison, Lutra lutra	Stuffed dummy	Rabbit
Olendorf and Robinson (2000)	At the nest	Mobbing	Empidonax virescens	Nature	Molothrus ater, Cyanocitta cristata	Stuffed dummy	None
Burhans (2001)	At the nest	Alarm calls	Spizella pusilla	Nature	Molothrus ater, Cyanocitta cristata	Stuffed dummy	Passerella iliaca
Gill and Sealy (2004)	At the nest	Mobbing calls, time at the nest	Setophaga petechia	Nature	Molothrus ater, Quiscalus quiscula	Stuffed dummy	None
Honza et al. (2004)	At the nest	Mobbing, latency, time at the nest	Acrocephalus scirpaceus	Nature	Cuculus canorus, human	Stuffed dummy, living	Coluba livia f. domestica
Roos and Pärt (2004)	At the nest	Nest selection	Lanius collurio	Nature	Corvus cornix, Pica pica	Living	Corvus monedula
Aviles and Parejo (2006)	At the nest	Mobbing index	Cyanopica cyanus, Pica pica	Nature	Accipiter nisus, Clamator glandarius	Stuffed dummy	Turdus viscivorus

(continued)

Table 1.3 (continued)

References	Method	Measured marker	Tested bird	Bird origin	Predators, parasites	Predator form	Control
Csermely et al. (2006)	At the nest	Mobbing, fleeing	*Falco tinnunculus*	Nature	*Corvus cornix, C. corax*	Stuffed dummy	None
D'Orazio and Neudorf (2008)	At the nest	Mobbing, calling	*Thryothorus ludovicianus*	Nature	*Molothrus ater, Cyanocitta cristata*	Freeze-dried dummy	*Catharus ustulatus*
Welbergen and Davies (2008)	At the nest	Distance, alarm calls	*Acrocephalus scirpaceus*	Nature	*Accipiter nisus, Cuculus canorus*	Stuffed dummy	*Anas crecca*
Campobello and Sealily (2010)	At the nest	Mobbing, calling	*Acrocephalus scirpaceus*	Nature	*Pica pica, Cuculus canorus*	Stuffed dummy	*Coluba livia f. domestica*
Strnad et al. (2012)	At the nest	Mobbing	*Lanius collurio*	Nature	*Pica pica, Garrulus glandarius, Accipiter nisus, Falco tinnunculus*	Stuffed dummy	*Columba livia f. domestica*
Thorogood and Davies (2012)	At the nest	Mobbing	*Acrocephalus scirpaceus*	Nature	*Accipiter nisus, Cuculus canorus* (two colour morphs)	Balsa dummy	None
Trnka and Prokop (2012)	At the nest	Mobbing	*Acrocephalus arundinaceus*	Nature	*Accipiter nisus, Cuculus canorus*	Stuffed dummy	*Streptopelia turtur*
Trnka et al. (2012)	At the nest	Mobbing	*Acrocephalus arundinaceus*	Nature	*Cuculus canorus*—with/without underpart barring and yellow eye	Stuffed dummy	*Streptopelia decaocto*
Trnka and Grim (2013)	At the nest	Mobbing	*Acrocephalus arundinaceus*	Nature	*Accipiter nisus, Cuculus canorus* (two colour morphs), *Falco tinnunculus*	Stuffed dummy	*Streptopelia turtur*
Liang and Møller (2015)	At the nest	Mobbing index	*Hirundo rustica*	Nature	*Accipiter nisus, Cuculus canorus*	Stuffed dummy	*Streptopelia orientalis*

Němec and Fuchs (2014)	At the nest	Mobbing	*Lanius collurio*	Nature	*Garrulus glandarius, Nucifraga caryocatactes, Corvus frugilegus, Corvus corone, Corvus corax*	Stuffed dummy	*Columba livia f. domestica*
Yang et al. (2014)	At the nest	Mobbing	*Prinia flaviventris*	Nature	*Cuculus optatus, C. canorus, Lanius schach, Garrulus glandarius*	Stuffed dummy	*Turdus pallidus, Streptopelia orientalis*
Yu et al. (2016)	At the nest	Various response behaviour—index	*Hirundo rustica*	Nature	*Accipiter nisus, Cuculus canorus*	Swallow alarm calls	*Streptopelia orientalis*
Syrová et al. (2016)	At the nest	Mobbing of *Falco tinnunculus*	*Lanius collurio*	Nature	*Garrulus glandarius, Pica pica*	Stuffed dummy	*Columba livia f. domestica*
Yu et al. (2017)	At the nest	Mobbing, alarm calling	*Parus major*	Nature	*Accipiter nisus, Cuculus canorus*	Stuffed dummy, calls	*Streptopelia orientalis*
Duré Ruiz et al. (2018)	At the nest	Alarm calling, feeding	*Troglodytes aedon*	Nature	*Milvago chimango, Buteo magnirostris, Harpagus bidentatus*	Stuffed dummy	*Chrysomus ruficapillus*
Strnadová et al. (2018)	At the nest	Mobbing	*Lanius collurio*	Nature	*Pica pica, Garrulus glandarius, Accipiter nisus, Falco tinnunculus*	Stuffed dummy	*Columba livia f. domestica*
Griesser (2009)	At the feeder	Mobbing calls	*Perisoreus infaustus*	Nature	*Accipiter gentilis, A. nisus, Falco subbuteo, Strix uralensis, Surnia ulula, Glaucidium passerinum*	Stuffed dummy	Empty control, *Cyanocitta cristata*

(continued)

Table 1.3 (continued)

References	Method	Measured marker	Tested bird	Bird origin	Predators, parasites	Predator form	Control
Soard and Ritchison (2009)	At the feeder	Mobbing calls, distance	*Poecile carolinensis*	Nature	*Megascops asio, Bubo virginianus, Accipiter striatus, A. cooperii, Buteo jamaicensis, Falco sparverius*	Stuffed dummies	Empty control, *Bonasa umbellus*
Courter and Ritchison (2010)	At the feeder	Mobbing calls	*Baeolophus bicolor*	Nature	*Megascops asio, Bubo virginianus, Accipiter striatus, A. cooperii, Buteo jamaicensis*	Stuffed dummies	Empty control, *Bonasa umbellus*
Tvardíková and Fuchs (2011)	At the feeder	Feeder attendance	*Parus major, Cyanistes caeruleus, Poecile* sp.	Nature	*Accipiter nisus, Falco tinnunculus*	Stuffed dummies	Empty control
Tvardíková and Fuchs (2012)	At the feeder	Feeder attendance	*Parus major, Cyanistes caeruleus, Poecile* sp.	Nature	*Accipiter nisus, Falco tinnunculus*	Stuffed dummies	*Coluba livia f. domestica, Turdus philomelos,* empty control
Trnka et al. (2015)	At the feeder	Arrival latency, flight frequency—index	*Passer montanus, P. domesticus*	Nature	*Accipiter nisus, Cuculus canorus* (two colour morphs), *Falco tinnunculus*	Stuffed dummy	*Streptopelia decaocto*
Carlson et al. (2017b)	At the feeder	Alarm calling	*Parus major, Cyanistes caeruleus*	Nature	*Accipiter nisus, Athene noctua*	Stuffed dummy	None

Reference	Location	Behaviour	Species tested	Setting	Predator species	Method	Control
Miller (1952)	In the field	Mobbing	Various bird species	Nature	*Bubo virginianus, Megascops asio, Glaucidium gnoma*	Call imitation	None
Altmann (1956)	In the field	Mobbing	Various bird species	Nature	*Megascops asio, Bubo virginianus, Athene cunicularia, Asio flammeus, Glaucidium gnoma*	Call imitation	None
Grubb (1977)	In the field	Alarm calls	*Fulica americana*	Nature	*Buteo lineatus, Haliaeetus leucocephalus, Pandion haliaetus*	Living	Aeroplane
Rasa (1981)	In the field	Mobbing, fleeing	*Tockus erythrorhynchus, T. deckeni, T. flavirostris*	Nature	*Accipiter badius, Butastur rufipennis, Falco tinnunculus, Falco naumanni*	Living	None
Adams et al. (2006)	In the field	Vigilance	*Platycercus elegans*	Nature	*Aquila audax, Falco peregrinus*	Playback	Conspecific
Reudink et al. (2007)	In the field	Mobbing	Various bird species	Nature	*Megascops asio, Glaucidium brasilianum*	Playback	None
Nocera and Ratcliffe (2009)	In the field	Mobbing	Various bird species	Nature	*Bubo virginianus, Aegolius acadicus*	Stuffed + playback alarm call of *Poecile atricapillus*	None
Da Cunha et al. (2017a)	In the field	Mobbing	19 bird species	Nature	*Glaucidium brasilianum, Athene cunicularia*	Stuffed + playback	None
Da Cunha et al. (2017b)	In the field	Mobbing	Various bird species	Nature	*Glaucidium brasilianum, Athene cunicularia*	Stuffed + playback	None
Nice and Ter Pelkwyk (1941)	Field observation	Alarm calls	*Melospiza melodia*	Nature	6 raptor species	Living	None

(continued)

Table 1.3 (continued)

References	Method	Measured marker	Tested bird	Bird origin	Predators, parasites	Predator form	Control
Kruuk (1964)	Field observation	Mobbing, fleeing	Chroicocephalus ridibundus	Nature	Circus spp., Falco peregrinus, Larus argentatus, Larus fuscus, Vulpes vulpes, Mustela erminea	Living	None
Gaddis (1980)	Field observation	Alarm calls	Various bird species	Nature	Accipiter striatus, A. cooperii, Buteo lineatus, Cathartes aura	Living	None
Buitron (1983)	Field observation	Mobbing	Pica pica	Nature	Falco mexicanus, F. columbarius, Buteo jamaicensis, Circus cyaneus, Accipiter cooperii	Living	None
Trail (1987)	Field observation	Fleeing, alarm calls	Rupicola rupicola	Nature	Spizaetus ornatus, Micrastur semitorquatus, Accipiter bicolor, Leucopternis albicollis, Buteogallus urubitinga, Morphnus guianensis	Living	Various harmless bird species
Walters (1990)	Field observation	Mobbing	Vanellus chilensis, V. armatus, V. crassirostris	Nature	Various bird, mammal and reptile species, human	Living	Various harmless animals
Winkler (1992)	Field observation	Mobbing	Tachycineta bicolor	Nature	Falco sparverius, Falco columbarius, Accipiter sp.	Living	None
Duckworth (1997)	Field observation	Mobbing	Macronous gularis	Nature	Surniculus lugubris, Dicrurus sp.	Living	None
Hafthorn (2000)	Field observation	Alarm calls	Poecile montanus	Nature	5 raptors, 2 owls, 5 corvids, 2 woodpeckers	Living	Various harmless bird species

Amat and Masero (2004)	Field observation	Mobbing, fleeing, crouching	*Charadrius alexandrinus*	Nature	*Falco tinnunculus, F. peregrinus, Milvus migrans, Hieraaetus pennatus, Circus pygargus, C. aeruginosus*	Living	None
Nijman (2004)	Field observation	Mobbing	*Dicrurus macrocercus, D. leucophaeus*	Nature	*Spizaetus bartelsi, Ictinaetus malayensis*	Living	None
Edelaar and Wright (2006)	Field observation	Alarm calls	*Turdoides squamiceps*	Nature	Various species of raptors, shrikes, gulls, corvids	Living	Various harmless bird species

and Temple 1986a, c, 1988; Winkler 1992, 1994) as well as species readily using nest boxes, the most frequent being the titmice (Curio et al. 1983; Curio and Regelmann 1985; Ficken et al. 1994; Curio and Onnebrink 1995), but also the flycatchers (Curio 1975; Shalter 1978; Dale et al. 1996; Bureš and Pavel 2003) and the nuthatches (Ghalambor and Martin 2000). Besides these, other species are also used, such as the thrushes (Shedd 1982; McLean et al. 1986; Hogstad 2005; Rodriguez-Prieto et al. 2009), the warblers (Edwards et al. 1950; Kleindorfer et al. 1996, 2005; Halupka 1999; Bureš and Pavel 2003) and different species of granivorous songbirds (Patterson et al. 1980). An aggressive form of defence of the trial species constitutes an advantage in the experiment evaluation (Edwards et al. 1950), Red-backed Shrikes being such a species in Central Europe (Ash 1970; Strnad et al. 2012; Němec and Fuchs 2014; Němec et al. 2015). It accomplishes quite high nesting densities in suitable biotopes (Roos and Pärt 2004; Goławski and Meissner 2007), while its nests are easy to find (Müller et al. 2005) and its physical disposition allows the bird to use the most aggressive forms of defence (Tryjanowski and Goławski 2004; Strnad et al. 2012). Similarly, the tyrant flycatchers (Tyrannidae), for instance, can be used on the North American continent (Blancher and Robertson 1982). Non-songbirds are used less frequently in nest experiments focused on predator recognition (Tables 1.1, 1.2, 1.3, 1.4 and 1.5), these primarily being colonially breeding species (Elliot 1985; Conover 1987; Burger and Gochfeld 1992; Clode et al. 2000; Stenhouse et al. 2005), whose advantages were already mentioned in connection with the observation approach to studying predator recognition. Nonetheless, there also have been some works concerning raptors (Arroyo et al. 2001; Csermely et al. 2006) and the owls (Hakkarainen and Korpimäki 1994).

1.2.2   Experiments on Winter Feeders

Providing birds with an attractive food source constitutes another situation when they can be made not to avoid the predator, winter feeders being such a source in cold and temperate climate zones. A feeder may represent a significant source of food, especially in the period when their natural sources are reduced (Lack 1954; Robb et al. 2008a, b). Visiting a feeder can then significantly contribute to the survival of an individual, thus increasing its fitness (Jansson et al. 1981). On the other hand, winter feeders also represent an interesting source of food for predators, which are sure to find a prey there (Dunn and Tessaglia 1994). Therefore, birds at the feeders face a permanent danger of predation, having to deal with a trade-off between the risk of predation and the use of food source when visiting the feeder (e.g. Abrahams and Dill 1989; Carrascal and Polo 1999; Carrascal and Alonso 2006).

Given that a predator is placed near a feeder, we bring about a situation when the incoming birds must decide whether to resign themselves to the food provided or whether to opt for undergoing the risk. The risk level can be adjusted by changing the distance of the predator from the feeder; however, the existing studies on

Table 1.4 Summary of studies testing the bird ability to recognize various predator species from different ecological guilds

References	Method	Measured marker	Tested bird	Bird origin	Ps	P form	Control
Schaller and Emlen (1962)	In the lab	Fleeing, vigilance—index	*Gallus gallus f. domestica, Phasianus colchicus torquatus*	Captive bred	*Megascops* sp., white rat	Latex dummy, living	Rooster, rectangle
Gyger et al. (1987)	In the lab	Alarm calls	*Gallus gallus f. domestica*	Captive bred—seminatural	Various aerial and ground predators	Living	Various harmless flying or ground objects
Evans et al. (1993a)	In the lab	Alarm calls	*Gallus gallus f. domestica*	Captive bred	Raptor-shaped image, *Procyon lotor*	Video	None
Kullberg (1998)	In the lab	Vigilance	*Poecile montanus*	Nature	*Accipiter nisus, Glaucidium passerinum*	Stuffed dummy	Empty control
Göth (2001)	In the lab	Vigilance, freezing, fleeing	*Alectura lathami*	Captive bred	Dog, cat, raptor silhouette, snake	Living, rubber, plywood dummy	Equally sized cardboard objects
Zaccaroni et al. (2007)	In the lab	Vigilance, freezing, fleeing	*Alectoris graeca*	Captive bred	Buzzard-like silhouette, *Vulpes vulpes*	Wooden, stuffed dummy	None
Binazzi et al. (2011)	In the lab	Vigilance, freezing, fleeing, alarm calls	*Alectoris rufa*	Captive bred	Buzzard-like silhouette, *Vulpes vulpes*	Wooden, stuffed dummy	None
Sieving et al. (2010)	In the lab	Alarm calls	*Baeolophus bicolor*	Nature	*Accipiter striatus, Felis catus, Megascops asio, Elaphe guttata*	Living, stuffed dummy	Empty cage
Dessborn et al. (2012)	In the lab	Vigilance, freezing, fleeing	*Anas platyrhynchos*	Captive bred	*Corvus corone, Larus canus, L. argentatus, Esox lucius, Accipiter gentilis, Neovison vison*	Playback, stuffed dummy,	*Chloris chloris, Cygnus cygnus, Fringilla coelebs*

(continued)

Table 1.4 (continued)

References	Method	Measured marker	Tested bird	Bird origin	Ps	P form	Control
Schetini de Azevedo et al. (2012)	In the lab	Vigilance, freezing, fleeing	*Rhea americana*	Captive bred	*Puma concolor, Cerdocyon thous, Buteo magnirostris*	living, moving dummy	Human, chair, *Tamandua tetradactyla*
Dutra et al. (2016)	In the lab	Index	*Sicalis flaveola*	Captive bred	*Rupornis magnirostris, Milvago chimachima*	Living, stuffed dummy	Lego cube
Azevedo et al. (2017)	In the lab	Index	*Amazona aestiva*	Captive bred	*Leopardus pardalis, Parabuteo unicinctus,* human	Stuffed dummy, living	Chair
Ash (1970)	At the nest	Mobbing	*Lanius collurio*	Nature	*Garrulus glandarius, Tyto alba, Strix nebulosa*	Stuffed dummy	None
Balda and Bateman (1973)	At the nest	Mobbing	*Gymnorhinus cyanocephalus*	Nature	*Bubo virginianus,* human	Stuffed dummy, living	None
Curio (1975)	At the nest	Mobbing	*Ficedula hypoleuca*	Nature	*Lanius collurio, Dendrocopos major, Glaucidium passerinum, Coracias garrulus*	Stuffed dummy	None
Gottfried (1979)	At the nest	Mobbing	*Cardinalis cardinalis, Turdus migratorius, Dumetella carolinensis, Toxostoma rufum*	Nature	*Cyanocitta cristata, Coluber constrictor*	Stuffed and plastic dummy	*Passer domesticus*
Patterson et al. (1980)	At the nest	Mobbing	*Zonotrichia leucophrys*	Nature	*Thamnophis elegans, Falco sparverius, Aphelocoma coerulescens*	Living, Freeze-dried dummy	*Junco hyemalis oreganus*
Blancher and Robertson (1982)	At the nest	Mobbing	*Tyrannus tyrannus*	Nature	Human, crow	Living, stuffed dummy	None

Shields (1984)	At the nest	Mobbing	*Hirundo rustica*	Nature	Human, *Megascops asio*	Living, stuffed dummy	None
Elliot (1985)	At the nest	Mobbing	*Vanellus vanellus*	Nature	*Vulpes vulpes, Corvus corone, Larus maritimus*	Stuffed dummy	*Columba palumbus*
Brown and Hoogland (1986)	At the nest	Mobbing	*Hirundo rustica, Petrochelidon pyrrhonota, Stelgidopteryx serripennis, Riparia riparia*	Nature	*Bubo virginianus, Mustela frenata*	Stuffed dummy	None
Knight and Temple (1986c)	At the nest	Mobbing	*Agelaius phoeniceus*	Nature	Human, *Procyon lotor*	Living, stuffed dummy	None
Mclean et al. (1986)	At the nest	Mobbing, alarm calls	*Turdus migratorius*	Nature	*Corvus brachyrhynchos*, human	Stuffed dummy, living	*Columba livia f. domestica*
Conover (1987)	At the nest	Mobbing	*Larus delawarensis*	Nature	Human, *Bubo virginianus*	Stuffed dummy, living	None
Knight and Temple (1988)	At the nest	Mobbing calls	*Agelaius phoeniceus*	Nature	*Procyon lotor, Buteo jamaicensis, Corvus brachyrhynchos*	Stuffed, plastic dummy	Paper box
Hauser and Wrangham (1990)	At the nest	Fleeing, vigilance	*Corythaeola cristata*	Nature	*Pan troglodytes, Stephanoaetus coronatus*	Playback	*Bycanistes subcylindricus, Piliocolobus badius*
Stone and Trost (1991)	At the nest	Alarm calls	*Pica hudsonia*	Nature	*Circus cyaneus, Corvus brachyrhynchos, Felis catus, Procyon lotor, Buteo jamaicensis, Accipiter gentilis*	Living	None
Burger and Gochfeld (1992)	At the nest	Mobbing	*Rynchops niger*	Nature	Human, *Larus argentatus*	Living	None

(continued)

Table 1.4 (continued)

References	Method	Measured marker	Tested bird	Bird origin	Ps	P form	Control
Winkler (1992)	At the nest	Mobbing	*Tachycineta bicolor*	Nature	*Mustela putorius furo, Elaphe obsoleta*	Living	None
Ficken et al. (1994)	At the nest	Alarm calls	*Parus sclateri*	Nature	*Glaucidium gnoma, Bubo virginianus*, human	Playback, plastic dummy, living	None
Winkler (1994)	At the nest	Mobbing	*Tachycineta bicolor*	Nature	*Mustela putorius furo, Pantherophis obsoletus*	Living	None
Curio and Onnebrink (1995)	At the nest	Distance	*Parus major*	Nature	*Dendrocopos major, Strix aluco*	Stuffed dummy	None
Dale et al. (1996)	At the nest	Return of parents at the nest	*Ficedula hypoleuca*	Nature	*Accipiter nisus, Dendrocopos major*	Stuffed dummy	*Turdus pilaris*
Kleindorfer et al. (1996)	At the nest	Hiding, fleeing	*Acrocephalus melanopogon*	Nature—chicks	Snake, human, *Circus aeruginosus*	Plastic dummy, living, stuffed dummy	None
Halupka (1999)	At the nest	Mobbing	*Acrocephalus paludicola*	Nature	*Circus cyaneus, Mustela putorius*	Stuffed dummy	*Emberiza schoeniclus*
Arnold (2000)	At the nest	Mobbing	*Manorina melanocephala*	Nature	*Falco berigora, Corvus orru*	Stuffed dummy	*Macropygia amboinensis*
Arroyo et al. (2001)	At the nest	Mobbing	*Circus pygargus*	Nature	*Corvus corone, Bubo bubo, Vulpes vulpes*	Stuffed dummy	None
Ghalambor and Martin (2000)	At the nest	Nest visits	*Sitta canadensis, S. carolinensis*	Nature	*Troglodytes aedon, Accipiter striatus*	Stuffed dummy	None

Bureš and Pavel (2003)	At the nest	Mobbing	*Ficedula hypoleuca, Sylvia atricapilla, Anthus pratensis*	Nature	*Garrulus glandarius, Corvus corax, Mustela erminea, Martes foina*	Stuffed dummy	None
Hogstad (2005)	At the nest	Mobbing	*Turdus pilaris*	Nature	*Corvus cornix, Accipiter nisus*, human	Stuffed dummy, Living,	None
Kleindorfer et al. (2005)	At the nest	Behavioural index	*Acrocephalus arundinaceus, A. scirpaceus, A. melanopogon*	Nature	Snake, *Mustela erminea, Circus aeruginosus*	Plastic, stuffed dummy	None
Stenhouse et al. (2005)	At the nest	Mobbing	*Xema sabini*	Nature	*Vulpes vulpes, Larus argentatus*	Stuffed, wooden dummy	*Histrionicus histrionicus*
Rodriguez-Prieto et al. (2009)	At the nest	Fleeing	*Turdus merula*	Nature-naïve, experienced	Familiar human, novel car toy	Living and model	None
Suzuki (2011)	At the nest	Mobbing calls	*Parus major*	Nature	*Corvus macrorhynchos, Elaphe climacophora*	Living	None
Królikowska et al. (2016)	At the nest	Mobbing	*Vanellus vanellus*	Nature	Various predators of eggs	Living	None
Bogrand et al. (2017)	At the nest	Alarm calling	*Thryothorus ludovicianus*	Nature	Snake, cat	Stuffed dummies	*Columba livia f. domestica*
Maziarz et al. (2018)	At the nest	Calls, leaving the nest, feeding chicks	*Phylloscopus sibilatrix*	Nature	*Garrulus glandarius, Mustela nivalis*	Stuffed dummy	Mug, nothing
Campobello and Sealy (2018)	At the nest	Distraction, mobbing, alarm calling	*Setophaga petechia, Acrocephalus scirpaceus*	Nature	*Molothrus ater, Cuculus canorus*	Living	None

(continued)

Table 1.4 (continued)

References	Method	Measured marker	Tested bird	Bird origin	Ps	P form	Control
Hauser and Caffrey (1994)	At the feeder	Fleeing	*Corvus brachyrhynchos*	Nature	*Bubo virginianus, Buteo lineatus, Polyboroides radiatus*—unfamiliar	Playback	*Hylocichla mustelina*
Griesser (2008, 2009)	At the feeder	Mobbing calls	*Perisoreus infaustus*	Nature	*Accipiter gentilis, Cyanocitta cristata*	Stuffed dummy	Empty control
Griesser and Suzuki (2017)	At the feeder	Mobbing callas	*Perisoreus infaustus*	Nature	*Accipiter gentilis, Accipiter nisus, Strix uralensis*	Stuffed dummy	Empty control
Naguib et al. (1999)	In the field	Mobbing calls	*Turdoides squamiceps*	Nature	*Asio flammeus, Felis catus*	Stuffed dummy, living	None
Rainey et al. (2004)	In the field	Mobbing calls	*Ceratogymna atrata*	Nature	*Stephanoaetus coronatus, Panthera pardus*	Playback	None
Stenhouse et al. (2005)	In the field	Mobbing	*Xema sabini*	Nature	*Vulpes vulpes, Larus argentatus*	Stuffed, carved dummy	*Histrionicus histrionicus*
Yorzinski and Vehrencamp (2009)	In the field	Mobbing	*Corvus brachyrhynchos*	Nature	*Bubo virginianus, Procyon lotor*	Stuffed dummy	None
Nice and Ter Pelkwyk (1941)	Field observation	Mobbing calls	*Melospiza melodia*	Nature	Raptors, owls, grackles, corvids, carnivores, snakes, human	Living	Rabbit, passerines
McNicholl (1973)	Field observation	Mobbing	*Sterna paradisaea, S. forsteri*	Nature	*Larus argentatus, L. delawarensis, Stercorarius parasiticus, Nycticorax nycticorax*	Living	None
Guillory and LeBlanc (1975)	Field observation	Mobbing	*Hirundo rustica*	Nature	*Egretta thula, Butorides virescens, Egretta tricolor, Tyrannus tyrannus, Agelaius phoeniceus, Quiscalus major*	Living	*Petrochelidon pyrrhonota, Riparia riparia*

Leger and Carroll (1981)	Field observation	Mobbing calls	Phainopepla nitens	Nature	Aphelocoma coerulescens, Buteo jamaicensis	Living	None
Siegel-Causey and Hunt (1981)	Field observation	Mobbing	Phalacrocorax pelagicus, P. auritus	Nature	Larus glaucescens, Corvus caurinus	Living	None
Brunton (1986)	Field observation	Injury-feigning	Charadrius vociferus	Nature	Vulpes vulpes, human	Living	None
Byrkjedal (1987)	Field observation	Fleeing, distraction	Charadrius morinellus, Pluvialis apricaria	Nature	Human, Vulpes vulpes, Corvus corax, Larus canus	Living	None
Brunton (1990)	Field observation	Crouching or fleeing	Charadrius vociferus	Nature	Human, carnivores, rodents, raptors, crows, gulls, grackles	Living	None
Green et al. (1990)	Field observation	Proportion of attacked intruders	Limosa limosa, Vanellus vanellus	Nature	Corvus corone, Falco tinnunculus, Ardea cinerea	Living	None
Jacobsen and Ugelvik (1992)	Field observation	Alert postures, freezing, fleeing	Anas penelope	Nature	Vulpes vulpes, Corvus corax, Corvus cornix, Accipiter gentilis, Aquila chrysaetos, Buteo lagopus, Bubo bubo	Living	None
Winkler (1992)	Field observation	Mobbing	Tachycineta bicolor	Nature	Mustela putorius, Pantherophis obsoletus	Living	None
Brunton (1997)	Field observation	Mobbing	Sternula antillarum	Nature	Nycticorax nycticorax, Corvus brachyrhynchos	Living	None
Amat and Masero (2004)	Field observation	Mobbing, fleeing, crouching	Charadrius alexandrinus	Nature	Dog, Falco tinnunculus, F. peregrinus, Milvus migrans, Hieraaetus pennatus, Circus pygargus, C. aeruginosus, Gelochelidon nilotica, Lanius excubitor, L. senator, Corvus corax	Living	None

(continued)

Table 1.4 (continued)

References	Method	Measured marker	Tested bird	Bird origin	Ps	P form	Control
Sordahl (2004)	Field observation	Mobbing	*Recurvirostra americana, Himantopus mexicanus*	Nature	Dog, fox, weasel, skunk, muskrat, harrier, kestrel, eagle, gull, tern, crane, raven, heron, owl	Living	None
Stenhouse et al. (2005)	Field observation	Mobbing	*Xema sabini*	Nature	*Alopex lagopus, Larus hyperboreus, Larus argentatus, Stercorarius parasiticus, Falco peregrinus*	Living	None
Palestis (2005)	Field observation	Mobbing	*Sterna hirundo*	Nature	Human, *Larus marinus*	Living	None
Murphy (2006)	Field observation	Wag-tail movements	*Eumomota superciliosa*	Nature	Human, *Urocyon cinereoargenteus*, dog, cat, *Nasua narica*, raptors, snakes, *Ctenosaura similis*	Living	Cattle, *Sylvilagus floridanus*

Table 1.5 Summary of studies testing the bird ability to recognize predators from non-predators

References	Method	Measured marker	Tested bird	Bird origin	Predators, parasites	Predator form	Control
Lorenz (1939)	In the lab	Fleeing, hiding	*Anser anser*	Captive bred	Raptor	Moving silhouette	Goose
Melzack et al. (1959)	In the lab	Fleeing, hiding, alarm calls	*Anas platyrhynchos f. domestica*	Captive bred	Raptor	Moving silhouette	Goose
Hinde (1960)	In the lab	Alarm calls	*Fringilla coelebs*	Wild-caught	*Strix aluco*	Stuffed dummy	Toy dog
Green et al. (1966)	In the lab	Activity level	*Anas platyrhynchos f. domestica*	Captive bred	Raptor	Moving silhouette	Goose, triangular shape
Green et al. (1968)	In the lab	Fleeing, hiding	*Anas platyrhynchos f. domestica*	Captive bred	Raptor	Moving silhouette	Goose
Melvin and Cloar (1969)	In the lab	Freezing	*Colinus virginianus*	Captive bred	*Buteo swainsoni*	Living	*Columba livia f. domestica*
Mueller and Parker (1980)	In the lab	Heart rate	*Anas platyrhynchos f. domestica*	Captive bred	Raptor	Moving silhouette	Goose
Moore and Mueller (1982)	In the lab	Heart rate	*Gallus gallus f. domestica*	Captive bred	Raptor	Moving silhouette	Goose
Canty and Gould (1995)	In the lab	Fleeing, hiding, alarm calls	*Anas platyrhynchos f. domestica*	Captive bred	Raptor	Moving silhouette	Goose
Hogstad (1995)	In the lab	Alarm calls	*Poecile montanus*	Wild-caught	*Accipiter nisus*	Stuffed dummy	*Picoides tridactylus*
Fluck et al. (1996)	In the lab	Contact with and distance from the P	*Gallus gallus f. domestica*	Captive bred	Cat	Scent	Neutral odour
Fransson and Weber (1997)	In the lab	Activity, food intake	*Sylvia atricapilla*	Wild-caught	*Accipiter nisus*	Stuffed dummy	Plastic bottle

(continued)

Table 1.5 (continued)

References	Method	Measured marker	Tested bird	Bird origin	Predators, parasites	Predator form	Control
Pravosudov and Grubb (1998)	In the lab	Vigilance, foraging, fat amount	*Baeolophus bicolor*	Wild-caught	*Accipiter striatus*	Stuffed dummy	*Zenaida macroura*
McLean et al. (1999)	In the lab	Mobbing	*Petroica australis*	Wild-caught-naïve	*Mustela furo*, cat	Stuffed dummy	Plastic bottle
Van der Veen (1999)	In the lab	Body mass changes	*Emberiza citrinella*	Wild-caught	*Accipiter nisus*	Stuffed dummy	Empty control
Bautista and Lane (2000)	In the lab	Mobility and physiology	*Periparus ater*	Wild-caught	*Strix aluco*	Playback	No sound, night jar calls
Canoine et al. (2002)	In the lab	Corticosteroid levels	*Saxicola torquata rubicola*	Wild-caught	*Strix aluco*	Living	Cloth bag, cage
Cockrem and Silverin (2002)	In the lab	Corticosterone levels	*Parus major*	Wild-caught	*Aegolius funereus*	Stuffed dummy	*Fringilla montifringila*, cardboard box
Kullberg and Lind (2002)	In the lab	Freezing	*Parus major*	Captive bred	*Accipiter nisus*	Stuffed dummy	*Perdix perdix*
Kullberg and Lind (2002)	In the lab	Freezing	*Parus major*	Wild-caught	*Accipiter nisus*	Stuffed dummy	*Perdix perdix*
Baker and Becker (2002)	In the lab	Alarm calls	*Parus atricapillus*	Wild-caught	*Falco mexicanus*	Stuffed dummy	Block of wood
Hagelin et al. (2003)	In the lab	Time spent at the odour	*Aethia cristatella*	Wild-caught	Mixture of mammalian scent with skunk smell	Scent	Banana essence
Lind et al. (2005)	In the lab	Mobbing calls	*Parus major*	Wild-caught	*Glaucidium passerinum*	Stuffed dummy	*Erithacus rubecula*

Cresswell et al. (2009)	In the lab	Vigilance, fleeing	*Fringilla coelebs*	Wild-caught	*Accipiter nisus*	Stuffed dummy	*Columba palumbus*
Gérard et al. (2015)	In the lab	Avoidance rating	*Puffinus pacificus*	Wild-caught	*Rattus rattus*	Scent	Water scent
Amo et al. (2015)	In the lab	Avoidance rating	*Carpodacus mexicanus*	Wild-caught	*Didelphis marsupialis*	Scent	Water scent
Griggio et al. (2016)	In the lab	Building nest	*Passer domesticus*	Captive bred	*Mus musculus domesticus*	Scent	Hay scent
McIvor et al. (2018)	In the lab	Flights	*Corvus monedula*	Wild-caught	*Vulpes vulpes*	Stuffed dummy	Toy elephant
Goodwin (1953)	At the nest	Mobbing	Various birds	Nature	*Garrulus glandarius*	Stuffed dummy	*Columba oenas*
Barash (1975)	At the nest	Fleeing	*Prunella collaris*	Nature	Human	Living	None
Fuchs (1977)	At the nest	Mobbing	*Sterna sandvicensis, S. hirundo, S. paradisaea, Chroicocephalus ridibundus*	Nature	*Larus argentatus*	Stuffed dummy	None
Andersson et al. (1980)	At the nest	Mobbing	*Turdus pilaris*	Nature	Human	Living	None
Bump (1986)	At the nest	Mobbing	*Xanthocephalus xanthocephalus*	Nature	*Cistothorus palustris*	Playback	Conspecific, *Melospiza melodia*, *Geothlypis trichas*
Hobson et al. (1988)	At the nest	Mobbing	*Setophaga petechia*	Nature	*Sciurus carolinensis*	Stuffed dummy	None
Onnebrink and Curio (1991)	At the nest	Mobbing	*Parus major*	Nature	*Dendrocopos major*	Stuffed dummy	None

(continued)

Table 1.5 (continued)

References	Method	Measured marker	Tested bird	Bird origin	Predators, parasites	Predator form	Control
Hakkarainen and Korpimäki (1994)	At the nest	Mobbing	*Aegolius funereus*	Nature	*Mustela vison*	Living	None
Maloney and McLean (1995)	At the nest	Mobbing	*Petroica australis*	Nature-naïve, experienced	*Mustela erminea*	Stuffed dummy	Plastic box
Halupka and Halupka (1997)	At the nest	Mobbing	*Anthus pratensis*	Nature	*Circus cyaneus*	Stuffed dummy	None
Hatch (1997)	At the nest	Distance from P, alarm calls	*Melospiza melodia*	Nature	*Corvus caurinus*	Stuffed dummy	*Junco hyemalis*
Nealen and Breitwisch (1997)	At the nest	Alarm calls	*Cardinalis cardinalis*	Nature	*Tamias striatus, Cyanocitta cristata, Elaphe obsoleta*	Stuffed or plastic dummy	*Zenaida macroura*
Meilvang et al. (1997)	At the nest	Nest defence behaviour	*Turdus pilaris, T. iliacus*	Nature	*Corvus corone*	Stuffed dummy	None
Michl et al. (2000)	At the nest	Return to the nest	*Ficedula hypoleuca*	Nature	*Accipiter nisus*	Stuffed dummy	*Turdus viscivorus*
Burhans (2000)	At the nest	Hiding	*Spizella pusilla*	Nature	*Cyanocitta cristata*	Stuffed dummy	*Pipilo erythrophthalmus*
Veen et al. (2000)	At the nest	Attack and mobbing calls	*Acrocephalus sechellensis*	Nature	*Foudia sechellarum, Passer domesticus* (female resembles F.s.),	Living, stuffed dummy	*Geopelia striata, Parus major*
Bosque and Molina (2002)	At the nest	Mobbing	*Cyanocorax cayanus*	Nature	Boa constrictor	Stuffed dummy	None

Hakkarainen et al. (2002)	At the nest	Food provisioning to chicks	Ficedula hypoleuca	Nature	Glaucidium passerinum	Stuffed dummy + playback	Turdus merula
Betts et al. (2005)	At the nest	Distance moved	Setophaga caerulescens, Setophaga virens	Nature	Poecile atricapillus mobbing at Aegolius acadicus	Playback	Conspecific song playbacks
Eggers et al. (2005)	At the nest	Nest visits	Perisoreus infaustus	Nature	Mixture of Garrulus glandarius, Corvus cornix and Corvus corax	Playback	No playback
Fisher and Wiebe (2006)	At the nest	Mobbing	Colaptes auratus	Nature	Tamiasiurus hudsonicus	Stuffed dummy	Xanthocephalus xanthocephalus, Bombycilla cedrorum
Godard et al. (2007)	At the nest	Willingness to breed in the nest box	Sialia sialis	Nature	Elaphe obsoleta, Peromyscus maniculatus	Scent	Neutral odour
Amo et al. (2008)	At the nest	Latency to enter the nest, time spent in it	Cyanistes caeruleus	Nature	Mustela putorius furo	Scent	Coturnix japonica scent; water 'scent'
Peluc et al. (2008)	At the nest	Nest position, food provisioning	Oreothlypis celata	Nature-naïve	Aphelocoma californica	Stuffed dummy + playback	Carpodacus mexicanus, empty control
Neudorf et al. (2011)	At the nest	Mobbing	Xanthocephalus xanthocephalus, Setophaga petechia	Nature	Cistothorus palustris, Troglodytes aedon	Freeze-dried dummy	Spizella pallida
Davidson et al. (2015)	At the nest	Return to the nest	Corvus monedula	Nature	Human with threatening mask	Living	Human with neutral mask
Wheatcroft et al. (2016)	At the nest	Mobbing, calling	Ficedula hypoleuca, F. albicollis	Nature	Accipiter nisus	Stuffed dummy	None

(continued)

Table 1.5 (continued)

References	Method	Measured marker	Tested bird	Bird origin	Predators, parasites	Predator form	Control
Chiver et al. (2017)	At the nest	Mobbing, killing	*Cymbilaimus lineatus*	Nature	*Pseustes poecilonotus*	Plastic model + alarm call of *C. lineatus*	None
Ha et al. (2018)	At the nest	Calling	*Parus minor*	Nature	Snake	Living	None
Gentle and Gosler (2001)	At the feeder	Amount of fat	*Parus major*	Nature	*Accipiter nisus*	Stuffed dummy	Plastic bottle
Cockrem and Silverin (2002)	At the feeder	Corticosterone levels	*Parus major*	Nature	*Aegolius funereus*	Stuffed dummy	None
Griffin et al. (2005)	At the feeder	Feeder attendance, excitement	*Zenaida aurita*	Nature	*Mustela erminea* with *Quiscalus lugubris* alarm call	Stuffed dummy	Cardboard box, *Quiscalus lugubris* song
Randler (2006)	In the field	Vigilance	*Fulica atra*	Nature	*Canis familiaris*	Playback	*Fringilla coelebs*
Scheuerlein et al. (2001)	Field observation	Corticosterone levels	*Saxicola torquata axillaris*	Nature	*Lanius collaris* presence	Living	*Lanius collaris* absence
Adriaensen et al. (1998)	Field observation	Body mass	*Cyanistes caeruleus*	Nature	*Accipiter nisus*	Living	Empty control

predator recognition placed it in the immediate vicinity (Hill 1986; Davies and Welbergen 2008; Tvardíková and Fuchs 2010, 2011, 2012) or no more than 1 m away (Soard and Ritchison 2009; Courter and Ritchison 2010), barring the studies of Griesser (2008, 2009), who placed mounted predators 5 m from the feeders visited by the Siberian Jays (*Perisoreus infaustus*).

At first glance, the experiments on feeders could seem to provide an ideal approach to the predator recognition research. Regrettably, they also bring considerable disadvantages. If in the case of nest experiments we are faced with their standardization, it is doubly true for feeder experiments.

A total anonymity of birds visiting the feeder constitutes the main problem. Most species using feeders create flocks that do not permanently dwell near the feeders, but linger there only on a one-time basis (Tvardíková and Fuchs 2010), this undoubtedly happening in order to reduce the risk of predation (Elgar 1989; Lima 1995; Beauchamp 2001). The frequency and regularity of visits varies with individual bird groups. The flocks of the tits visit feeders relatively often and regularly (Jansson et al. 1981; Shedd 1983; Hinsley et al. 1995; Koivula et al. 1995; Kullberg 1998; Pravosudov and Grubb 1998; Krams 2000, 2001; Gentle and Gosler 2001; Krams et al. 2006), while the intervals between the visits of granivorous bird flocks (Fringillidae, Emberizidae) are longer and irregular—probably because they can ingest larger amounts of food (Lilliendahl 1997; Quinn and Cresswell 2005; Roth et al. 2008). Feeders are also visited by species permanently dwelling in the vicinity, as they have more or less permanent winter territories and areas, in Central Europe these being, e.g. the robins, blackbirds, nuthatches and woodpeckers (Tvardíková and Fuchs 2012), as well as the willow tit (*Poecile montanus*) and crested tit (*Lophophanes cristatus*) in Sweden (Jansson et al. 1981). However, they usually occur at the feeder individually; therefore, the changes in their numbers do not provide enough information on how they evaluate the risk of predation. The uneven use of feeders over time can be compensated in two ways. Either the predator presentation is very short (5 min—Davies and Welbergen 2008; Soard and Ritchison 2009; Courter and Ritchison 2010), thus affecting a homogeneous sample of the bird population attending the vicinity of the feeder at the time, or on the other hand, the predator can be presented for so long that a large proportion of the total bird population manages to visit the feeder (30 min in the studies of Griesser 2008, 2009; Tvardíková and Fuchs 2010, 2011, 2012).

Some cases deal with the issue of the anonymity of birds visiting the feeder, e.g. studies on the Siberian Jay (Griesser 2008, 2009), where the researchers had at their disposal a caught and colour-coded group of birds whose mutual social relationships were well known (Ekman et al. 2001; Griesser 2003, 2008, 2009). These works, along with the work of Hauser and Caffrey (1994) concerning the American Crow (*Corvus brachyrhynchos*), are the only ones that did not use the tits as study subjects in researching predator recognition on the feeder (Tables 1.1–1.3). In their case, the problem of anonymity is frequent.

In fact, the constitution of tit flocks varies over time—some individuals leave while others join (Tvardíková and Fuchs 2010). The modification speed is different in different species. Most European tits, except for the crested tit and willow tit, which are very sedentary during the winter (Jansson et al. 1981), create anonymous flocks with a low stability (Gaddis 1980). As far as the North American species are concerned, the same goes for the black-capped chickadee (*Poecile atricapillus*—Shedd 1983; Apel 1985; Smith 1997; Gunn et al. 2000), the Carolina chickadee (*Poecile carolinensis*—Freeberg and Lucas 2002; Soard and Ritchison 2009), the Mexican chickadee (*Poecile sclateri*—Ficken et al. 1994) and the tufted titmice (*Baeolophus bicolor*—Hill 1986; Pravosudov and Grubb 1998; Courter and Ritchison 2010). The stability of flocks breeds a pseudoreplication problem. If the flocks were absolutely stable throughout the whole winter, we should not repeat experiments on one feeder, or more precisely on one locality. However, this is difficult to accomplish, not only for technical reasons (see below). At least in the case of species forming anonymous flocks, the composition modification does occur. Its speed can be determined by repeatedly catching the birds and their subsequent individual ringing. Quinn et al. (2012) enriched this method with PIT (passive integrated transponder) tagging of the caught tits, the tags being read by loggers placed next to individual feeders thus enabling the quantification of the number of visiting birds. Taking into consideration that the series of experiments usually take a few days to carry out (Tvardíková and Fuchs 2010), the flock stability and consequent pseudoreplication do not constitute a fatal problem; however, they should be taken into account when evaluating the results.

As in the case of the nest experiments, the experiments on feeders are also affected by the environment, the accessibility of hiding places being absolutely essential. If birds visiting a feeder have an option of a quick escape into hiding, their willingness to take risks significantly increases. In one study of Tvardíková and Fuchs (2012), 31% of the tits were willing to risk perching on a feeder with the sparrowhawk in the vicinity, while in another study (Tvardíková and Fuchs 2011), it was a maximum of 5% of the tits. In the first case, the experimental feeder was surrounded by bushes from three sides no further than at a distance of 4 m, whereas in the second one, it was located on a meadow 7 m away from a bushy baulk. The major effect of a hiding place availability on the willingness of birds visiting the feeder to risk results in the impossibility to compare the behaviour towards the same species (modification) of a predator between individual locations (feeders), but only the differences in behaviour towards different species (modifications). The frequency and regularity of visits to the feeders is obviously also affected by the supply of other food sources nearby. A greater variability in responses of the tits to the same predator observed in towns would suggest that, since winter supplementary feeding of birds is widespread there (Fuller et al. 2008; Davies et al. 2012), as opposed to rural landscape (Fuchs personal observation). Quite unsurprisingly, the willingness to take risks is significantly influenced by the weather as well. In conditions of very low temperatures and a substantial snow blanket, the motivation of birds to visit the feeder increases, despite the presence of a predator, with their willingness to take risks rising (Tvardíková and Fuchs 2011; Fig. 1.1). Therefore, these parameters must always be included in the data analysis.

The preceding paragraphs clearly imply that the absolute numbers of arrivals at a feeder are affected by many external factors; therefore, they cannot be used as a

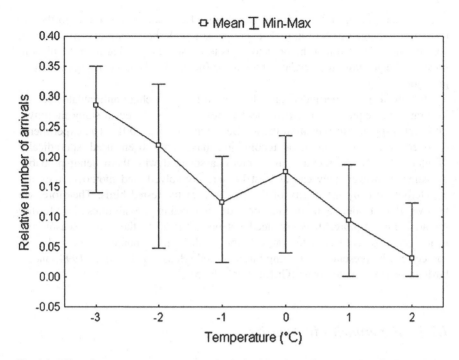

Fig. 1.1   Effect of temperature on the number of arrivals of four tit species to the winter feeder—relative number of arrivals to the feeder with predator compared to the control feeder without any dummy (Tvardíková and Fuchs 2011 with permission)

reliable benchmark of the birds' willingness to take risks, and they must be related to some kind of control. The simplest way to get the control is to alternate trials where the feeder is freely accessible with trials with a predator placed at the feeder. The drop in the number of arrivals at the feeder after exposing the predator constitutes the benchmark of the birds' willingness to risk. However, not even this way of data correction can eliminate short-term fluctuations in the movement of individual flocks, as they can, for instance, omit some of the regularly visited localities. The optimal solution would be if the control took place simultaneously with the exposure of the predator. A 'two-feeder' design allows that (Tvardíková and Fuchs 2010, 2011). The birds present are offered two feeders located within sight of each other. The predator subject to trial is placed on them, while the control species is placed on the other. An extremely dangerous predator (the sparrowhawk in Central Europe) or a harmless bird of a similar size as the predator (e.g. the domestic pigeon in Central Europe) can be the control species. Provided that more birds come to the predator than to the sparrowhawk during the experiment, the trial predator causes a lesser fear than the sparrowhawk. The tits might have recognized a dangerous bird in it, however not the sparrowhawk. This comparison is important if we modify the potential recognition features of the sparrowhawk (whether those shared with other predators or species-specific). If fewer birds come to the trial predator than to the pigeon, the trial predator causes a bigger fear than the pigeon; ergo the tits recognize a dangerous bird in it. The comparison with the pigeon is then used as a control one. Provided that the predator fails to cause a bigger fear than the pigeon, none of the features the predator has are sufficient for it to be recognized as a predator.

Hitherto feeder experiments have been used for researching antipredator behaviour rather infrequently—most frequently they have been used to study the physiological responses to predation threat (Gentle and Gosler 2001). The experiments on feeders to study predator recognition have only been used sporadically (Tables 1.1–1.3), the methodical problems associated with them being probably to blame. However, they can at least be partially solved, and moreover they are counterbalanced by the clarity of the response of the tested birds. Therefore, we believe that feeder experiments provide interesting possibilities for further research. Furthermore, their use need not be restricted to the winter season and to temperate or cold climate zones of the northern hemisphere. Additional food sources can be presented to nesting birds as well (Hauser and Caffrey 1994) and to birds in warm climate zones (Griffin et al. 2005).

1.2.3   Experiments in Aviaries

Antipredator behaviour of birds can also be researched in captivity. This approach, unlike field experiments, allows standardizing experimental conditions. The main disadvantage is that these conditions are fundamentally different from those in the wild. If the trial birds have been captured in the wild, their main source of stress most probably lies in the captivity—even if they are habituated to their new environment (Van Dongen et al. 2001). The tits in cage experiments spend most of their time exploring, undoubtedly driven by their desire to escape. They do pay attention to the predator present; however, it is lesser than in the natural experiments (cf. Beránková et al. 2015; Veselý et al. 2016).

The existing studies researching antipredator behaviour using experiments that take place in captivity have had a long tradition. Entirely aviary experiments have been used in pioneer studies aimed at finding features the birds are guided by when recognizing predators (Lorenz 1939; Krätzig 1940; Nice and Ter Pelkwyk 1941; Melzack et al. 1959; Hinde 1960). This approach enjoyed great development from the 1960s (Schaller and Emlen 1962; Melvin and Cloar 1969; Scaife 1976; Klump and Curio 1983; Evans et al. 1993a, b; Flasskamp 1994; Kullberg 1998), especially when studying the recognition of raptorial flight silhouettes (Green et al. 1968; Mueller and Parker 1980; Moore and Mueller 1982; Canty and Gould 1995). Studies on alarm calls (Bautista and Lane 2000; Dessborn et al. 2012) and on the recognition of odours (Fluck et al. 1996; Hagelin et al. 2003) are also connected with predator recognition. In recent years, cage experiments have experienced a significant renaissance, being the most common method when studying predator recognition after experiments carried out on nests (Lind et al. 2005; Palleroni et al. 2005; Templeton et al. 2005; Azevedo and Young 2006a, b; van den Hout et al. 2006, 2010a, b; Zaccaroni et al. 2007; Carter et al. 2008; Cresswell et al. 2009; Mathot et al. 2009; Sieving et al. 2010; Binazzi et al. 2011; Schleidt et al. 2011; Dessborn et al. 2012; Schetini de Azevedo et al. 2012; Beránková et al. 2014, 2015). Naturally, the organization of experiments varies from study to study. At first, only responses of birds to randomly occurring wild predators were observed (Gyger et al. 1987), nonetheless soon to be followed by experiments using various forms of simulated presentation of predators.

The older studies in particular worked with domesticated species (chicken, duck, goose, turkey) or those bred in captivity (willow grouse, *Lagopus lagopus*; northern bobwhite, *Colinus virginianus*; partridge, *Alectoris*; song sparrow, *Melospiza melodia*) (Table 1.1). However, this limits their informative value, since the birds seem to gain a significant part of their knowledge about predators by learning (Curio 1975; Maloney and McLean 1995; McLean et al. 1999; Kullberg and Lind 2002). On top of that, the domesticated species tend to reduce their antipredator behaviour conduct, relying on human protection to a degree (Schaller and Emlen 1962). The domestication may eventually alter their cognitive abilities as well as their behavioural repertoire (e.g. Kagawa et al. 2014).

Studies where captive-bred birds are presented with predators in order to create an antipredator behaviour of an endangered species before its release into the wild represent a special case (Hölzer et al. 1996; McLean et al. 1999; Wong 1999; Göth 2001; Azevedo and Young 2006a, b; Schetini de Azevedo et al. 2012). Out of the birds captured in the wild, positively the most common species tested in captivity were the titmice (Klump and Curio 1983; Alatalo and Helle 1990; Hogstad 1995; Kullberg 1998; Pravosudov and Grubb 1998; van der Veen 1999; Bautista and Lane 2000; Baker and Becker 2002; Kullberg and Lind 2002; Lind et al. 2005; Templeton et al. 2005; Sieving et al. 2010; Beránková et al. 2014, 2015). Besides these, the species sometimes used also included the thrushes and starlings (Flasskamp 1994; Carter et al. 2008), the warblers (Fransson and Weber 1997), granivores (Hinde 1960; Cresswell et al. 2009) and the waders (Hagelin et al. 2003; van den Hout et al. 2006, 2010a, b; Mathot et al. 2009). The stress resulting from captivity is probably so high with many species that they are unable to express normal antipredatory behaviour (Dickens and Bentley 2014). Therefore, the tits appear to be clearly the most appropriate group in that respect. Even in the case of adaptable species, a certain amount of time to adapt must be taken into account. The birds must not only become accustomed to captivity as such but also get acquainted with the experimental aviary or cage. The great tits (*Parus major*) tested by Beránková et al. (2014) were held captive for 1–3 days before the trial, being placed into the experimental cage 10 min before the experiment. This organization proved to be sufficient, as the birds expressed a rich repertoire of antipredator and comfort behaviour that seem to match their natural behaviour well.

1.2.4   Presentation of Predators

The experimental approach to studying predator recognition breeds a question—in what form and way should the predators be presented to the tested birds. Live predators are not used so much (Tables 1.1–1.3). Besides the difficulty in obtaining them, the inconveniences associated with their presentation are probably the reason, as the presentation must be credible, but at the same time being capable of preventing a direct contact of the predator with the tested birds. This can be ensured by a presentation in a cage (Curio et al. 1983; Curio and Regelmann 1985; Flasskamp 1994; Sieving et al. 2010); however, a question arises to what extent the birds sense that the predator in the cage poses a real threat. Fastening the predator to the perch appears more suitable (Curio et al. 1983). In addition to that, the predator behaviour should be similar in all trials. These requirements can only be met with big difficulties, particularly in the case of birds of prey. Nevertheless, there have been some studies succeeding in it. Palleroni et al. (2005) used three representatives of the Accipiter genus in their experiments. Trained individuals kept overflying an aviary with hens in it, which showed varied antipredator behaviour with respect to the level of perceived risk. Templeton et al. (2005) even presented 13 species of the owls and birds of prey to the black-capped chickadees sitting on a perch near the chickadees' home aviary. Besides birds, live dogs appear in several studies as well (Göth 2001; Azevedo and Young 2006a, b), so do cats (Naguib et al. 1999; Göth 2001; Templeton et al. 2005; Sieving et al. 2010), minks (Hakkarainen and Korpimäki 1994; Dessborn et al. 2012), ferrets (Winkler 1992, 1994; Templeton et al. 2005) and rats (Schaller and Emlen 1962), and even live snakes (Suzuki 2011).

Nonetheless, the vast majority of experiments carried out not only within the research of predator recognition but also of other aspects of antipredator behaviour used the predator decoys instead of live predators. In the case of bird predators, these usually were mounted individuals in the upright-seated position, where the absence of movement did not seem unnatural (Table 1.1), the exceptions being studies using mounted birds flying with wings spread, mostly drawn on the wire (e.g. Gentle and Gosler 2001; van den Hout et al. 2006, 2010a, b; Mathot et al. 2009). Numerous studies on mammalian predators worked with mounted animals as well (Table 1.1).

Mounted animals are fully suited for researching the ability of birds to recognize predators. However, the research of the actual recognition processes requires manipulating with features which the presented predators have in order to be recognized. The possibilities of the manipulation are rather limited though, especially regarding the mounted birds, yet they have been used for this purpose (Edwards et al. 1950; Curio 1975; Scaife 1976; Smith and Graves 1978; Gill et al. 1997a; Davies and Welbergen 2008; Tvardíková and Fuchs 2010; Trnka et al. 2012). Replacing mounted animals with models seems to be an optimum solution (Tables 1.1 and 1.2), as they i.a. resolve the poor availability of quality mounted animals. That was probably why models were used in studies where individual features were not manipulated with (Table 1.1). On the other hand, using models breeds a problem with their credibility. The aforementioned studies mostly failed to address it, being content with the trial birds usually responding to the presented models. Several works have compared to what extent mounted animals successfully replace live predators (Curio 1975; Shalter 1978; East 1981; Blancher and Roberson 1982; Shedd 1982; Knight and Temple 1986b; Meilvang et al. 1997); however, the testing of the models' effectiveness was only carried out once (Němec et al. 2015). Comparing mounted animals and models made of textile and plastic showed that the reduced credibility can affect the reaction of the tested birds, in fact to a significant extent. Special attention should be primarily paid to the surface material. Using plastic models in the case of snakes seem to cause the least problems due to their easily imitable body surface (Patterson et al. 1980; Kleindorfer et al. 1996, 2005; Göth 2001). The presentation of printed still images or videos using a monitor constitutes, of course, another method of presentation in laboratory conditions (Krätzig 1940; Nice and Ter Pelkwyk 1941; Evans et al. 1993a, b). This method shall be further dealt with in Chap. 4.

As already mentioned, the absence of movement is the common deficiency of most used mounted animals and models, this especially being noticeable in the case of mammalian and reptilian predators (Bažant 2009; Dessborn et al. 2012). Ergo, some studies have provided the presented predators with movement (Frankenberg 1981; Gentle and Gosler 2001; Cockrem and Silverin 2002; van den Hout et al. 2006, 2010a, b; Mathot et al. 2009; Dessborn et al. 2012).

1.3 Predator Recognition Markers

If we examine students' knowledge of predators (and other animals for that matter), we need not deal with the issue of which markers indicate that a student recognized them, as we created more or less generally accepted names for individual species and higher taxa, which we require from the students. However, that does not mean that a student who cannot name, e.g. the kestrel, does not really know it, as the bird could have nested on a balcony of the student's dorm room, where he/she regularly saw it, nonetheless not bothering to look up its conventional name.

We also find ourselves in the examiner's situation when we try to determine whether the trial birds know the predators we are presenting them with. Except for one partial exception that will be described later, we do not have a generally accepted set of names we could ask the birds about. Therefore, we must try to get an answer from their reactions during their encounter with the predator. This reaction may either be behavioural or physiological. We assume that if the tested bird recognizes the predator, it will react to it, and the form of this reaction will depend on the size and nature of the danger the predator poses for the bird (McLean and Rhodes 1991; Caro 2005). However, these assumptions are not self-evident, which must be constantly kept in mind. If the trial bird evaluates the existing danger as insignificant, it will not react at all. In the case the bird evaluates the danger impending from various predators as the same its response to them will not differ. Nonetheless, it does not mean that the bird did not recognize the predators in either of the two cases.

This i.a. implies that the more alike the danger impending from presented predators is, the lesser chance we have to prove that the tested birds distinguish them. It is thus evident that the following applies more inexorably to researching predator recognition than to other issues dealt with through behavioural sciences—if you fail to falsify the null hypotheses, you cannot draw any conclusions from it.

1.3.1 Behavioural Markers

The vast majority of studies on predator recognition use a behavioural response of the trial birds to real or simulated dangers (Tables 1.1 and 1.2), this undoubtedly happening for methodical reasons, since our possibilities to notice a physiological response are very limited, particularly in field experiments (changes in fat reserves of the great tits depending on the exposure to stress of a predator present on the feeder—Gentle and Gosler 2001).

As already pointed several times, avoiding an encounter with the predator constitutes a preferred antipredator behaviour (Caro 2005). In most types of experiments, we try to get birds into a situation where they fail to use this option—either because they give preference to defending their nests or obtaining food or because we physically prevent that. In the last case, the tested bird has no other choice than coping with the proximity of a predator. Nevertheless, this may lead to the fact that its behaviour will not be adequate to the impending danger, as the bird is not ready for such a situation (Van Dongen et al. 2001).

However, the effort to avoid predators is often a pursued marker in laboratory experiments. Considering the fact that the species of *Galliformes* and *Anseriformes* are the ones often tested, their common response to the predator is by so-called freezing (Krätzig 1940; Melvin and Cloar 1969; Klump and Curio 1983; Göth 2001; Kullberg and Lind 2002; Zaccaroni et al. 2007; Binazzi et al. 2011; Dessborn et al. 2012; Schetini de Azevedo et al. 2012), or by trying to hide in a hiding place (Lorenz 1939; Krätzig 1940; Melzack et al. 1959; Green et al. 1968; Canty and Gould 1995; Palleroni et al. 2005) or by just simply running away (Lorenz 1939; Melzack et al. 1959; Schaller and Emlen 1962; Green et al. 1968; Canty and Gould 1995; Göth 2001; Palleroni et al. 2005; Zaccaroni et al. 2007; Cresswell et al. 2009; Binazzi et al. 2011; Dessborn et al. 2012; Schetini de Azevedo et al. 2012) and staying away from the predator as far as possible (Scaife 1976). Admittedly, the efforts to avoid the predator were also observed in the cases of nest defence (Barash 1975; Hauser and Wrangham 1990; Kleindorfer et al. 1996; Burhans 2000; Csermely et al. 2006; Rodriguez-Prieto et al. 2009) and the feeder experiments (Hauser and Caffrey 1994; Soard and Ritchison 2009).

Naturally, the trial birds must on the whole show an active defence in the nest and feeder experiments. In the case of feeder experiments, birds deal with a trade-off between the risk of predation and the gain from the food offered. As already mentioned before, the level of risk and gain is influenced by many factors (Abrahams and Dill 1989), which the birds take into account when making a decision, or should take into account; still, the result is unequivocal at a level of an individual—it either

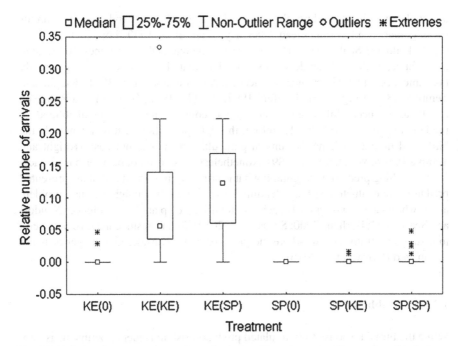

Fig. 1.2 Relative changes on the number of arrivals of four tit species (the number of arrivals to feeder with dummy/the number of arrivals to feeder during the reference control) to kestrel and sparrowhawk in trials with different treatment on alternative feeder (noted in parenthesis). Treatments on both feeders: KE, kestrel; SP, sparrowhawk; 0, empty feeder (Tvardíková and Fuchs 2011 with permission)

visits the feeder or does not. Therefore, if the bird's visits to the feeder with a predator decrease compared to the control, it means the bird has recognized the predator. Where the number of visits to the feeders with various predators differs, it means that the birds distinguished between them. Nonetheless, one question remains, and it concerns the birds that visited the feeder. Their visit to the feeder can have two reasons—they either decided to take the risk or they failed to recognize the predator, which is quite a big difference, if you are interested in their cognitive abilities. As shown by Tvardíková and Fuchs (2011), even this question can be answered by an appropriately structured experiment. In the case when the tits had a choice between a feeder with the common kestrel (*Falco tinnunculus*) and a feeder with the pigeon, the number of birds flying towards the kestrel was marginal, nonetheless multiplicatively increasing when the Eurasian sparrowhawk (*Accipiter nisus*) was placed on the second feeder instead of the pigeon. This rise can univocally be attributed to birds that knew the kestrel, flying near it; only then it represented a smaller risk (Fig. 1.2).

The choice of birds in nest experiments is simple at first glance, as well. They deal with the trade-off between the risk of predation and the gain from saving the offspring (Cordero and Senar 1990; Sordahl 1990). Nevertheless, there is actually a double risk in defending the nests—for the offspring and for the parents (Montgomerie and Weatherhead 1988; Rytkönen et al. 1990; Forbes et al. 1994; Campobello and Sealy 2010). The first of them increases the willingness to invest in the defence, while the latter decreases it. Besides that, the defending birds apparently take into account their chance to succeed, too (Lemmetyinen 1971; Knight and Temple 1986c; Burger and Gochfeld 1992; Winkler 1992; Olendorf and Robinson 2000), as unsuccessful defence can on the contrary increase the likelihood of predation (Gill et al. 1997b). Therefore, the right prediction of how intensively the birds will defend their nest against a particular predator is not easy (Knight and Temple 1986b; Weatherhead 1989). Nonetheless, this need not be a serious obstacle for researching predator recognition; what is primarily important is that the presented predators are evaluated by the defending birds differently, though a certain problem arises when the defending birds decide to totally give up an active defence (Neudorf and Sealy 1992; Burhans 2000; Strnad et al. 2012). Such a result can be very difficult to distinguish from a result where the predator is not considered a danger or is not recognized (Syrová et al. 2016).

1.3.1.1   Mobbing

While the birds' response to a simulated predation risk in feeder experiments is very simple, the antipredatory behaviour employed when defending nests is characterized by a high level of complexity. In literature, this became known as 'mobbing' (Hartley 1950). The term includes all behavioural elements whose function is to discourage a predator from attacking the nest but also to warn conspecifics as well as heterospecifics located near the place of the encounter, especially chicks in the nests (Curio et al. 1978; Templeton and Greene 2007; Magrath et al. 2010). Mobbing has many forms ranging from a mere presence of the parents near the predator (Curio and Onnebrink 1995; Deppe et al. 2003; Hogstad 2005; Welbergen and Davies 2008, 2011) through emitting alarm calls (Knight and Temple 1988; Stone and Trost 1991; Bazin and Sealy 1993; Ficken et al. 1994; Nealen and Breitwisch 1997; Burhans 2001) up to physical attacks (Shields 1984; Winkler 1994; Strnad et al. 2012; Němec and Fuchs 2014) and diverting the attention from the nest (Gochfeld 1984; Byrkjedal 1987; Sordahl 1990).

Mobbing is usually divided into passive and active (Curio 1976; Harvey and Greenwood 1978; Conover 1987; Caro 2005). Passive mobbing involves the parents remaining close to the predator and vocalization, ergo activities where the undertaken risk is relatively small. Staying close to the predator is accompanied by specific body postures, such as ruffling feathers, moving a tail or wings and squatting down (Ash 1970; Kumar 2003; Griffin et al. 2005). Passive mobbing can also include forming small flocks and flying in circles around the predator (Shields 1984; Conover 1987). Active mobbing comprises of physical harassment and attacking the predator (Shields 1984; Winkler 1994; Strnad et al. 2012; Němec and Fuchs 2014), in which the defending parents undergo a high risk of injury or death (Brunton 1986; Curio and Regelmann 1986; Poiani and Yorke 1989; Sordahl

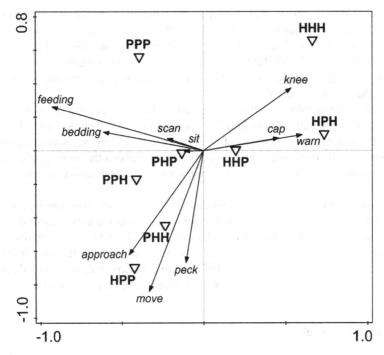

Fig. 1.3 Behaviour of great tits in the presence of individual dummies visualized on the first and second axis of PCA (Principle Component Analysis). The position of average reaction to the various chimeras of sparrowhawk and pigeon. The eye, beak and body features are represented by the first, second and third alphabets; P, pigeon; H, sparrowhawk (Beránková et al. 2014 with permission)

1990). On the other hand, if they are physically well-equipped, such active mobbing can be very effective.

Most behavioural elements associated with mobbing are quite distinct, thus being fairly easily measurable. In the case of active mobbing, the number of active attacks carried out by parents can be quantified. There are several degrees of active mobbing, which can be evaluated separately, ranging from approaching a defined distance through flying over the predator up to flying towards the predator while descending or directly attacking it physically (Ash 1970; Curio 1975; Kruuk 1976; Shields 1984; McLean et al. 1986). In the case of passive mobbing, what is measured is the time the parents spend at a distance from the predator and the number of performed warnings and typical attitudes or flyovers near the predator (Curio and Onnebrink 1995; Deppe et al. 2003; Hogstad 2005; Welbergen and Davies 2008, 2011). Passive mobbing constitutes a more complex behaviour than active mobbing; therefore, indexes tend to be used in its assessment (Schaller and Emlen 1962; Kleindorfer et al. 2005; Avilés and Parejo 2006; Honza et al. 2006; Liang and Møller 2015), so are multivariate statistical methods, such as the principal component analysis (Beránková et al. 2014; Fig. 1.3), in order to reduce the number of analysed variables. In both types of mobbing, it may be useful to distinguish its intensity, which can be measured by the total number of attacks or a total time of interest, from hazardousness, which reflects the degree of aggressiveness of the attacks or the average distance from a predator. For example, the shrikes in experiments of Strnad et al. (2012) attacked most of the presented predators with similar intensity, the aggressiveness of the attacks decreasing with the growing hazardousness of the predators, though.

Ruffling, alarm calls, squatting down and shaking wings could be observed in cage experiments with the tits (Kullberg and Lind 2002; Beránková et al. 2014, 2015). Besides these, the elements of exploratory and foraging behaviour commonly occurred as well. Therefore, a principal component analysis had to be employed for the assessment, showing three basic attitudes towards a predator—fear, interest and indifference.

Passive mobbing can also be noticed in feeder experiments. Nevertheless, so far it has not been used to evaluate the behaviour of trial birds towards presented predators, which is probably caused by the fact that activities taking place in relatively wide surroundings of the feeders cannot be easily noted. Furthermore, mobbing around the feeders is less frequent or less intense than expected, since a food source fails to represent such a strong subject of interest for the birds as the nest.

1.3.1.2 Alarm Calls

As already indicated, one of the functions of mobbing is warning the conspecifics as well as heterospecifics located near the place of the encounter (Curio 1976). This is done by a specific group of vocal signals called alarm calls (Naguib et al. 2009). Obviously, their efficacy will increase if they contain some information about the nature of the impending danger. Thus, it is not entirely true that birds cannot provide us with direct information on how they distinguish between predators they encounter, the accuracy of this information only being the question.

It has been repeatedly proven in experiments that birds distinguish in their alarm calls large groups of predators, differing in their hunting abilities, specifically birds vs. snakes (Sieving et al. 2010; Suzuki 2011, 2012) and land vs. aerial predators (Knight and Temple 1988; Walters 1990; Kleindorfer et al. 1996). The information about individual groups of predators can be passed on by completely different signals (Knight and Temple 1988; Griesser 2008, 2009; Sieving et al. 2010), but more often by changes of quantitative parameters of a commonly used basic signal. These may include different numbers of alarm calls emitted per unit of time (Evans et al. 1993a; Burhans 2001; Rainey et al. 2004), different numbers of individual syllables in a single alarm call (Walters 1990; Naguib et al. 2009; Binazzi et al. 2011) or even a different length and frequency of the individual syllables (Stone and Trost 1991; Ficken et al. 1994).

Species forming non-anonymous societies most probably pass on tremendously detailed information in their alarm calls. The first pieces of evidence come from the semiarid Arabian babblers (Naguib et al. 1999). However, this issue was studied the most elaborately on the North American black-capped chickadees, which emit two main types of alarm calls: a soft 'seet' used when a predator is flying near and a broad-band 'chick-a-dee', warning about a sitting predator. The first call makes the other chickadees hide in vegetation, while they respond to the second one by joint mobbing the sitting predator (Ficken et al. 1978; Smith 1997; Templeton and Greene 2007). The variability of the second type of the alarm call is a specialty of the black-capped chickadees. The more relevant and intense danger the predator poses, the more 'chick' syllables occur in the call. The drawn-out final syllable 'dee' elicits more intense group mobbing (Baker and Becker 2002), while the number of the 'dee' syllables indicates the size of the predator (Templeton et al. 2005). Faster repeating of the whole call 'chick-a-dee' warns of the predator being closer (Baker and Becker 2002). The association of the extraordinary complexity of the black-capped chickadee's alarm calls with its sociality supports a much poorer alarm call register of those titmice forming only volatile anonymous flocks (Lind et al. 2005; Courter and Ritchison 2010).

The existing knowledge undoubtedly proves that the alarm calls provide more or less detailed direct information on whether birds recognize predators they encounter. Nonetheless, at the same time, it points out one rigour complicating the interpretation of this information. Alarm calls give information not only about the predator itself but also on the circumstances of its occurrence, such as the distance (Templeton et al. 2005). As far as recognizing the predator is concerned, not even alarm calls thus represent the ideal markers that could be given preference over others.

1.3.2   Physiological Markers

An encounter with a predator undoubtedly represents a potential stressful event for its potential prey, the stress affecting a number of physiological parameters (Orchinik 1998). Therefore, the physiological markers would appear to be predetermined for studying the responses of birds to the encounters with predators. Besides it can be assumed that being a direct response of the organism to a stressful stimulus, they will be less affected by other circumstances (Scheuerlein et al. 2001).

Yet during the research of predator recognition, they were relatively rarely used (Tables 1.1–1.3). The reasons are obvious, lying in hard-to-solve methodical problems. Indirect markers of stress-strain have mostly been used in the research of predator recognition hitherto, e.g. body weight (Adriansen et al. 1998; van der Veen 1999; Bautista and Lane 2000; van den Hout et al. 2006, 2010a, b) or the formation of pectoral muscles (van den Hout et al. 2006, 2010a, b) and fat reserves (Pravosudov and Grubb 1998; Gentle and Gosler 2001; Scheuerlein et al. 2001).

As far as the directly measurable signs of the organism's stress-strain are concerned, heart rate measurement appears to be best usable (reviewed by Butler et al. 2004). This method was used on birds quite a long time ago (Owen 1969). In order to carry out this method in field experiments currently prevailing in the research of predator recognition, the heart rate measuring transducers would have to meet demanding requirements, especially small dimensions (as the scientists primarily work with songbirds), which is currently the main complication. Earlier studies used a small transducer affixed to the bird's chest, which increased the volume of the heart rate (Mueller and Parker 1980; Moore and Mueller 1982), while later ECG transmitters implanted into the bird's abdominal cavity were used, both in laboratory (McPhail and Jones 1998) and field conditions (Ely et al. 1999; Ackerman et al. 2004). However, a surgical procedure poses a significant risk, at the very least influencing the natural behaviour of the bird.

Newly, ECG connected to a Tattletale Lite computer in a harness attached to the back of the bird can be used—again in laboratory and field conditions (Enstipp et al. 1999; Storch et al. 1999). Nevertheless, this device partially restricts the natural behaviour, but above all, it is only suitable for larger birds (cormorants). Recently, an elegant non-invasive method applicable to small songbirds has been developed. The heartbeat monitor is in this case placed in fake eggs, measuring the heartbeat of incubating birds (Arnold et al. 2011). Naturally, this method cannot be used in predator recognition, either. However, the currently available mass-produced transducers intended for experiments with small mammals (Gilson and Kraitchman 2007) are not convenient, since they presume movement in a confined space, or even only on a monitoring mat (Harkin et al. 2002). The future probably lies in using an ECG monitor clipped to the surface of the bird's body (Johnson-Delaney 2003). Nonetheless, its conversion into miniscule forms that can be used in the fieldwork is still difficult.

Measuring the levels of stress hormones provides an attractive option (Cook 2012; Johnstone et al. 2012). In the last 20 years, the determination of their metabolites from faeces has been developing quite quickly (Lane 2006). Regardless of the most likely eliminable method accuracy problems such as fluctuations throughout the day (Carere et al. 2003) and the season (Astheimer et al. 1994), but also the dependence on the sex, food and individuality (Goymann 2012), the question remains whether the method is suitable for capturing short-term stress events, including the encounters with a predator. The relationship between the stress intensity and the metabolite level is not necessarily so simple (e.g. compare Chávez-Zicchinelli et al. 2014 and Albano et al. 2015). The possibilities of determining the stress hormones from feathers have also not been quite explored (Jenni-Eiermann et al. 2015). Nevertheless, it indeed is possible to determine a long-term or repeated stress-strain, which also brings some results more or less related to the topic of antipredator behaviour (Ylönen et al. 2006).

A direct determination represents a second option (Breuner et al. 2013). Besides the natural variability of individuals (Cockrem 2013), a very fast response to a stressful stimulus constitutes the main source of problems here (Deviche et al. 2012). The hormones need to be determined as quickly as possible while avoiding their level being affected by the actual sample taking (Van den Hout et al. 2010b). This can hardly be accomplished in field experiments, not being so simple in the laboratory, either (Canoine et al. 2002). Therefore, the levels of stress hormones were used in field research primarily in the situation where the birds were exposed to a predator in the long-term. Scheuerlein et al. (2001) ascertained that the tropical African stonechats (*Saxicola torquata*) nesting in the predator's territory have higher levels of corticosteroids in their plasma. Similarly, Clinchy et al. (2011) showed that the song sparrows have higher concentrations of glucocorticoids, provided that a higher predation pressure has been noted in their territory. Some studies (Dufty and Crandall 2005; Tilgar et al. 2010) also demonstrated that chicks have elevated levels of stress hormones if alarm calls are played for them for a certain period of time. The only study testing the influence of a predator on the immediate corticosteroid level in the blood was carried out by Canoine et al. (2002). Hand-raised stonechats were confronted with three situations of stress-strain. The results showed that the immediate corticosteroid level in the plasma (within 30 min after the treatment) was highest when the birds were confronted with the live tawny owl (*Strix aluco*), while it was lower when they were placed in a cage, and even lower when they were left in a dark bag.

Despite all the aforementioned obstacles, using physiological markers to research antipredator behaviour of birds in the broadest meaning still undoubtedly represents an attractive way. Above all, it would allow dealing with a situation when a trial bird surrenders the obvious behavioural response to predation risk.

Chapter 2
Evidence for Abilities of Predator Recognition

If we study the ability of birds to recognize predators, we must first make clear how accurate recognition we are interested in. It is the same as if we were to survey children's ability to recognize vehicles. They can recognize vehicles intrinsically: passenger cars vs. lorries, passenger Volkswagen vs. Mercedes cars and even Volkswagen Passat vs. Volkswagen Golf. We should try to specify a similar level of accuracy even when dealing with predator recognition. However, as already mentioned in Chap. 1, we cannot, naturally, ask whether birds recognize, e.g. Volkswagen and Mercedes cars, but only if they distinguish them.

Undoubtedly, distinguishing predators from harmless animals represents a basic level of the ability to recognize predators. To specify the subsequent levels, different criteria are possible. At first glance, it would be logical to follow the similarity in appearance of the predators whose distinguishing we are studying. However, we will run into a problem of how to quantify the similarity in appearance since current zoology almost does not address this issue, with some exceptions, such as the study of aposematism (e.g. measuring colour reflectance—Cibulková et al. 2014). In addition to that, the similarity in appearance may not even be a criterion for the potential prey. As already mentioned in Chap. 1, it cannot be assumed that the tested birds will behave differently to predators that differ in hunting skills or food preferences.

It is the differences in the degree of predator dangerousness given by their hunting skills and food preferences that come into consideration as an alternative criterion, since they should motivate the potential prey to distinguish between predators. Furthermore, we can determine the dangerousness of individual species of predators fairly accurately, as the data regarding their hunting methods and food are available, albeit of varying quality. However, the same problem will occur as in the case of similarity. The differences in the dangerousness of predators do not reflect the demands associated with their distinguishing. At least in theory, we can also imagine cryptic species (morphologically undistinguishable—Borkin et al. 2004), whose dangerousness will vary significantly.

R. Fuchs et al., *Predator Recognition in Birds*, SpringerBriefs in Animal Sciences, https://doi.org/10.1007/978-3-030-12404-5_2

Ergo, it is obvious that when evaluating the existing knowledge on the ability of birds to recognize or distinguish predators, we will have to take into account both criteria, asking to what extent the distinction is functional and to what extent it is difficult, limiting ourselves to four levels: (1) distinguishing predators and harmless animals, (2) distinguishing ground and aerial predators, (3) distinguishing aerial predators of adult birds and nests and (4) distinguishing different species of predators differing in hunting abilities and food preferences.

2.1 Predator vs. Non-predator

The ability to distinguish a predator from a harmless animal is a necessary precondition for effective antipredator behaviour, which can significantly increase the fitness of the attacked individual (Caro 2005). Unsurprisingly, this ability was ascertained not only in birds (McLean and Rhodes 1991) and mammals (McLean et al. 1996) but also in numerous invertebrates (Doherty and Hoy 1985). From a total of 37 studies testing the birds' ability to distinguish predators from harmless animals (as well as other objects) (Table 1.5), 34 of them proved it in one form or another. They most often worked with visual stimuli, whether with mounted animals, live animals or models (Table 1.5). The interest in acoustic stimuli was much smaller (Bump 1986; Bautista and Lane 2000; Hakkarainen et al. 2002; Betts et al. 2005; Eggers et al. 2005; Randler 2006; Peluc et al. 2008), while olfactory stimuli were for obvious reasons tested only rarely (Fluck et al. 1996; Hagelin et al. 2003; Godard et al. 2007; Amo et al. 2008, 2015; Gérard et al. 2015)—nonetheless, with a positive outcome as well.

The question of the birds' ability to distinguish predators from harmless animals could seem to be answered quite clearly and not worth being concerned with. However, this would be true only in the event that distinguishing a predator from a harmless animal was equally difficult, which certainly is not the case. The difficulty of the task is undoubtedly affected by the level of the predator appearance dissimilarity in particular. To distinguish the Eurasian sparrowhawk from the pigeon is obviously easier than distinguishing the Eurasian jay (*Garrulus glandarius*). However, there is more to it—the motivation to distinguish constitutes the second side of the matter. The motivation has two elements, the first one being the danger of a predator, predetermined by its hunting abilities and food preferences, and the second one being the likelihood of an encounter with it, predetermined especially by its abundance as well as biotope demands. Indisputably, the great tit is more motivated to distinguish a specialized predator of small birds—the European sparrowhawk—than a specialized predator of small mammals, the common kestrel. Similarly, the Central European great tit is motivated more to distinguish the commonly occurring European kestrel than the very rare red-footed falcon (*Falco vespertinus*).

Experiments carried out with birds bred in captivity without the possibility of learning to recognize predators suggest that birds seem to have some innate awareness of the existence and form of potential predators (Lorenz 1939; Nice and Ter Pelkwyk 1941; Melzack et al. 1959; Green et al. 1968; Melvin and Cloar 1969;

Scaife 1976; Mueller and Parker 1980; Moore and Mueller 1982; Evans et al. 1993a, b; Canty and Gould 1995; Fluck et al. 1996; Göth 2001; Zaccaroni et al. 2007; Binazzi et al. 2011; Dessborn et al. 2012). This conclusion is also confirmed by some studies focused on antipredatory reactions of birds from isolated islands, which have never met predators, either (Maloney and McLean 1995; McLean et al. 1999; Peluc et al. 2008). However, this knowledge is obviously not detailed enough, being probably refined during the life by individual or social acquisition. Individual experience is another factor affecting the ability of birds to distinguish predators from harmless animals. Unfortunately, these topics are neglected in the research of bird ability of predator recognition.

The birds of prey and owls were the most commonly tested bird predators (Table 1.5), specifically the northern goshawk (*Accipiter gentilis*) and the European kestrel from the first group and the pygmy owl (*Glaucidium*), great horned owl (*Bubo virginianus*) and eastern screech owl (*Megascops asio*) from the other.

The diversity of species used was much wider among the animals that served as harmless control, different Columbidae being the most frequent choice, i.e.—the pigeons (*Columba livia* f. domestica, *C. oenas*, *C. palumbus*) and doves (*Streptopelia turtur*, *S. decaocto*, *S. orientalis*). In addition to these, all kinds of the thrushes were often used as well (*Turdus philomelos*, *T. merula*, *T. pallidus*, *T. viscivorus*, *T. pilaris*, Table 1.5). Birds of prey and owls are both characterized by a relatively uniform appearance and the presence of distinct and more or less unique features (hooked beak, claws, eye ridge, veil). Ergo, it is not surprising that the trial birds distinguished predators and the control in nearly all studies.

Being one of the most significant bird predators of nests worldwide, the representatives of the Corvidae family (*Corvus cornix*, *C. corone*, *C. corax*, *C. frugilegus*, *C. brachyrhynchos*, *C. macrorhynchos*, *C. caurinus*, *Pica pica*, *Cyanocitta cristata*, Table 1.5) were the second most frequently tested group. As opposed to the owls and birds of prey, the Corvidae family is quite variable in appearance, and moreover, it lacks unique characteristic features, perhaps except for a huge beak. On the other hand, different species usually have distinct and well distinguishable colouration. Even though the recognition of the Corvidae family represents a difficult task at first sight, the trial birds managed to distinguish them not only from the control but also individual corvids from one another depending on their dangerousness.

Brood parasites also constitute an interesting group in terms of recognition. In spite of not being predators, their impact on the fitness of the hosts is, nevertheless, at least comparable (Rothstein 1990). Brood parasites are not endowed with any general features that would differentiate them from harmless birds; therefore, their recognition should be difficult. In addition to that, individual acquisition can only be made use of in a limited way, since the encounter with the parasite is not directly followed by its consequences. Active defence against the brood parasite may moreover be often ineffective or even counter-productive; on the other hand, identifying and removing an 'alien' egg is undemanding as far as time and energy are concerned, and if own eggs are distinguished properly from the alien ones, it can also constitute a very effective method of defending against the brood parasite (Rothstein 1990). In spite of this, there are plenty of studies dealing with the issue of brood parasite recognition (and not only its eggs). The common cuckoo (*Cuculus canorus*) and the brown-headed cowbird (*Molothrus ater*) are the most frequent types of nest predators whose recognition has been tested.

The existing knowledge on ground predator recognition is much less extensive than on the aerial predators (Tables 1.3–1.5). Quite unsurprisingly, the representatives of more or less specialized groups, i.e. carnivorans and snakes (Tables 1.3–1.5), have been the most frequently tested predators. They are fundamentally different in the variability of their appearance on the one hand and in the presence of features enabling their distinguishing from harmless animals on the other. While the appearance of snakes is extraordinarily uniform and also quite different from other ground vertebrates—should we ignore legless lizards—carnivorans constitute a relatively heterogeneous group, and moreover they lack distinct unique features that would make them different as a whole from harmless mammals.

Regrettably, most studies dealing with the recognition of carnivorans and snakes neglect a control in their design, and the capability to recognize a predator is usually inferred from the presence or absence of active or passive mobbing (Table 1.5). If a control is used in the experiment, it is usually either a harmless bird species (Patterson et al. 1980; Leger and Carroll 1981; Elliot 1985; Nealen and Breitwisch 1997; Halupka 1999; Griffin et al. 2005; Stenhouse et al. 2005; Randler 2006; Peluc et al. 2008; Dessborn et al. 2012) or an inanimate object (Schaller and Emlen 1962; Knight and Temple 1988; Maloney and McLean 1995; McLean et al. 1999; Göth 2001; Griffin et al. 2005; Azevedo and Young 2006a, b; Schetini de Azevedo et al. 2012). Only rarely did some mammals serve as the control, namely, the rabbit (Nice and Ter Pelkwyk 1941; Clode et al. 2000; Murphy 2006; Schetini de Azevedo et al. 2012), cattle (Murphy 2006) and anteater (Schetini de Azevedo et al. 2012). However, such an approach does not allow clarifying whether the trial birds actually recognize carnivorans or snakes, or whether they negatively respond to anything remotely resembling carnivorans or snakes.

While it is almost solely carnivorans from all the mammals that hunt adult birds, nests—those placed on the ground in particular—are endangered by the representatives of other groups as well, especially rodents and even insectivores. Nonetheless, the ability to recognize them has rarely been tested. Specifically, there have been two studies. Schaller and Emlen (1962) found out that chickens bred in captivity fail to recognize a danger in a white laboratory rat. Hobson et al. (1988) discovered that the American yellow warbler (*Setophaga petechia*) attacked a mounted eastern grey squirrel (*Sciurus carolinensis*) placed near the nest.

In the case of mammals, we can test predator recognition as a visual stimulus as well as a possible reaction of birds to its smell. Such studies were carried out both in laboratory conditions (Fluck et al. 1996; Hagelin et al. 2003; Amo et al. 2015; Gérard et al. 2015) and also on the nest (Godard et al. 2007; Amo et al. 2008). Nevertheless, the results of these studies are rather ambiguous. The crested auklets (*Aethia cristatella*) eschewed the smell of the skunk (Hagelin et al. 2003), while the house finches (*Carpodacus mexicanus*) avoided the smell of the common opossum (*Didelphis marsupialis*—Amo et al. 2015), and the blue tits (*Cyanistes caeruleus*) reacted to odour of the ferret (*Mustela putorius furo*) near the nest (Amo et al. 2008). However, only some chickens reacted to the cat smell (Fluck et al. 1996), while the wedge-tailed shearwater (*Puffinus pacificus*) did not evade the smell of the rat (Gérard et al. 2015), and contrary to presumptions, the eastern bluebirds (*Sialia sialis*) did not avoid nesting boxes with the smell of the deer mouse (*Peromyscus maniculatus*—Godard et al. 2007).

As already mentioned, the birds' ability to recognize predators is at the least improved by acquisition. Yet there are only few studies where identical predators were presented to inexperienced and experienced individuals of the same species. By a vast majority, these studies researched the ability of various poultry (chicken, duck, turkey) to recognize a predator in a raptor silhouette (rev. in Schleidt et al. 2011). However, these studies always compared birds bred in indoor farming with birds from outdoor aviaries, whose experience with predators was at best very limited. Moreover, it is rather their angular size and velocity than their shape that appears to be more important. The study of Kullberg and Lind (2002) is the only study to our knowledge, which really deals with the comparisons of responses of birds bred in captivity with birds caught wild. The ability to distinguish between a mounted sparrowhawk and a mounted partridge (*Perdix perdix*) was researched in this study. While inexperienced tits aged 30 days responded to both mounted birds in the same way, the tits caught in the wild (about 4 months old) distinguished between them.

Experiments carried out with birds that have lived without any contact with predators for a long time, e.g. on isolated islands, to some extent suggest the necessity of learning to distinguish between a predator and non-predator. For instance, it has been ascertained that the New Zealand robin (*Petroica australis*) living on a small island cannot distinguish between the weasel and a cardboard box of the same size. In contrast the robins from the mainland (the main New Zealand islands) where the weasels commonly occur responded to the stimuli presented in a different way (Maloney and McLean 1995).

Inexperienced birds have been used in many other studies, however without being compared with experienced birds, these specifically being the chicken (Schaller and Emlen 1962; Scaife 1976), the greater rhea (*Rhea americana*—Azevedo and Young 2006a, b; Schetini de Azevedo et al. 2012), the Australian brushturkey (*Alectura lathami*—Göth 2001), the mallard duck (*Anas platyrhynchos*—Dessborn et al. 2012) and the northern bobwhite (Melvin and Cloar 1969). Nonetheless, the results of these studies have been considerably inconsistent. The chickens were afraid of the kestrel (Scaife 1976), but failed to fear the screech owl (Schaller and Emlen 1962). The rhea did not distinguish between the control and different types of mammalian predators (Azevedo and Young 2006a, b; Schetini de Azevedo et al. 2012). In contrast, the ducks and bobwhites did recognize predators (Melvin and Cloar 1969; Dessborn et al. 2012). The brushturkeys were the best of the aforementioned species in predator recognition. Not only did they distinguish between a predator and a harmless object, but they responded to different kinds of predators in a specific way in accordance with the type of danger the predators represented (Göth 2001). The reason for their success in recognizing predators without any prior experience with them is probably to be found in their ontogenesis. The megapodes are extremely precocial, therefore having to rely only on their own innate abilities immediately after hatching. Unlike other species of birds, they cannot count on their parents' protection or the opportunities to learn from them.

Besides the choice of a species tested and its experience, the control used can significantly affect the result of the experiment. Using an abundant harmless animal (bird) size-wise comparable with the tested predators appears to be most certain. In this case, the risk that the presented control will cause fear in the trial animal, whether due to its excessive size or, e.g. owing to neophobia, is decreased. Using a so-called empty control is relatively frequent (Tables 1.3–1.5). Responding to the predator in that case can be compared with the bird's behaviour in the absence of any stimulus. Such an experimental arrangement is not entirely optimal, though, not enabling to eliminate a generalized antipredatory response to any 'intruder', which noticeably reduces the evidential value of such experiments. Using inanimate—usually artificial—objects also appears problematic. In the case of raptor silhouettes, various geometric shapes of a corresponding size serve as the control (Krätzig 1940; Nice and Ter Pelkwyk 1941; Schleidt et al. 2011). In other experiments, various objects were used as the control: a plastic bottle (Maloney and McLean 1995; Fransson and Weber 1997; McLean et al. 1999; Gentle and Gosler 2001), paper or plastic boxes (Knight and Temple 1988; Maloney and McLean 1995; Griffin et al. 2005), chairs (Azevedo and Young 2006a, b; Schetini de Azevedo et al. 2012), paper geometric shapes (Göth 2001), a wooden prism (Baker and Becker 2002) or a child's toy (Hinde 1960). There are two kinds of danger here. The objects can be entirely uninteresting for the trial birds; ergo they will respond to them less than they would react to a harmless bird. Nor we cannot rule out that a new noticeable object placed, for instance, near the nest will cause a neophobic reaction. Sordahl (2004) found out that birds had approached unfamiliar objects placed near the nest with a certain degree of cautiousness, nonetheless never attacking them, and their presence did not prevent the nesting pair from visiting the nest.

Using a harmless bird species—but unknown for the trial birds (Tables 1.3–1.5)—constitutes an interesting, although rarely used, possibility. Such a form of control was used, for example, by Welbergen and Davies (2009) in their study, who presented a mounted cuckoo and a same-sized balsa wood model of the generalized parrot (dark green upperparts, pale-green underparts) to the Eurasian reed warblers (*Acrocephalus scirpaceus*). Furthermore, Veen et al. (2000) used a mounted great tit as the control in his work on the Seychelles warblers (*Acrocephalus sechellensis*), which is actually an existing bird, yet the tested birds had never had a chance to encounter it. If the birds tested in these studies had responded to both the predator and the control by a similar antipredator behaviour, it would have suggested their ability to distinguish predators from harmless animals as well as unknown animals from known ones. The category of predators, or rather dangerous animals, would thus have been defined not positively (those being dangerous), but negatively (those being not harmless). However, no such outcome has been confirmed in the studies carried out so far, and it seems that birds tend to have a general idea what to be afraid of, ignoring unknown stimuli to a certain extent.

2.2   Ground vs. Aerial Predators

As already indicated, the research on the ability to recognize predators has especially been focused on bird predators, which implies that only a small number of studies have dealt with distinguishing aerial and ground predators (Table 1.4). This concerns experimental studies in particular. The most common comparison tested has been a response to raptors and carnivorans. The reaction was studied under laboratory conditions (Evans et al. 1993a; Göth 2001; Zaccaroni et al. 2007; Sieving et al. 2010; Binazzi et al. 2011; Dessborn et al. 2012; Schetini de Azevedo et al. 2012), on nests (Brown and Hoogland 1986; Knight and Temple 1988; Halupka 1999; Arroyo et al. 2001; Kleindorfer et al. 2005) or in the field (Naguib et al. 1999; Rainey et al. 2004; Yorzinski and Vehrencamp 2009). Sporadic studies tested the ability to distinguish between carnivorans and corvids or the seagulls (Elliot 1985; Bureš and Pavel 2003; Stenhouse et al. 2005); one study focused on distinguishing between a raptor and a snake (Kleindorfer et al. 1996); and one study even researched the distinguishing between the crow and the snake (Suzuki 2011). The results of these studies confirm the ability of the tested birds to distinguish between aerial and ground predators, with some exceptions (Brown and Hoogland 1986; Schetini de Azevedo et al. 2012). Observational studies are also in accord with experimental studies (Byrkjedal 1987; Brunton 1990; Jacobsen and Ugelvik 1992; Amat and Masero 2004; Sordahl 2004; Murphy 2006).

Nevertheless, the analyses of alarm and mobbing calls provide the most pieces of evidence concerning the ability of birds to distinguish between aerial and ground predators (Knight and Temple 1988; Stone and Trost 1991; Evans et al. 1993a; Naguib et al. 1999; Rainey et al. 2004; Suzuki 2011), as they have repeatedly demonstrated that alarm calls against these two types of predators patently differ from each other. In the presence of aerial and ground predators, birds either emit alarm calls with a varying intensity (Evans et al. 1993a; Naguib et al. 1999; Rainey et al. 2004) and length (Stone and Trost 1991) or emit totally different alarm calls (Knight and Temple 1988; Suzuki 2011).

Still this begs a question whether birds actually distinguish specific predators of these categories or only the direction of an impending danger. The methodology applied to most studies dealing with the question of recognizing aerial and ground predators fails to enable determining the cause of a difference in the response, as it tests both options concurrently. The aerial predator is usually either presented only as a flying silhouette or is placed perching on a tree stand. In contrast, ground predators are always logically placed on the ground. The hypothesis that the response depends more on the direction the danger is coming from rather than on the type of a predator was supported by one of the first studies on this topic (Gyger et al. 1987), which analysed the vocalizations of the domestic fowl cocks kept in outdoor aviaries. What was observed was the vocalization in response to a variety of naturally occurring objects in the vicinity of the aviary. Ground alarm calls were a typical reaction to any objects moving on the ground while objects moving in the air elicited aerial alarm calls.

The results of Kleindorfer et al. (2005) are also interesting. They found out that the reaction of three closely related Acrocephalus warblers (*A. melanopogon, A. scirpaceus* and *A. arundinaceus*) depends on the position of their own nest. Pairs who had a nest near the ground showed a stronger response against ground predators (snake; stoat—*Mustela erminea*). In contrast, birds with high-placed nests were more aggressive towards aerial predators (western marsh harrier—*Circus aeruginosus*). Consequently, the authors infer that birds not only distinguish ground and aerial predators, but they even take account of their current dangerousness in their response.

The identification of alarm calls holds a potential source of erroneous interpretations. Palleroni et al. (2005) show that the classification of alarms calls in 'ground' and 'aerial' calls is probably factitious to a large extent and that in fact these alarm calls may have much more information encoded in them than just the direction of a danger. In Palleroni's experiments, the males and females of the domestic fowl emitted aerial alarm calls more frequently in response to a large predator, whereas they used the ground alarm call to react to rather smaller predators. Likewise, Templeton et al. (2005) tested the vocal response of the black-capped chickadees to different-sized ground and aerial predators. The alarm calls regarding individual predators significantly differed in the number of syllables, with the size of the predator—but not its type—being the explanatory variable.

2.3   Nest vs. Adult Predators

Distinguishing bird predators of nests from bird predators of adult birds have been researched more frequently than distinguishing ground and aerial predators. This probably is due to the fact that this ability can significantly increase the fitness of nesting birds; ergo it has been of a considerable interest to behavioural ecology (Table 1.4). Regarding bird predators, raptors and the owls usually pose a risk for parents and weaned chicks, while eggs and chicks in the nest are especially endangered by corvids, along with the woodpeckers in the case of birds nesting in tree cavities. Therefore, both groups have most frequently been used in experiments.

The results of observational and experimental studies (Table 1.4) are also in accord with the assumption about distinguishing the predators of nests and adult birds being favourable. Birds usually react more intensely to specialized nest predators—it is manifested by their attacking them more often and with a greater vigour (Ash 1970; Curio 1975; Gottfried 1979; Patterson et al. 1980; Elliot 1985; Halupka 1999; Nijman 2004; Sordahl 2004; Hogstad 2005; Strnad et al. 2012). Smaller danger the specialized nest predators pose to adult birds results in their greater willingness to take the risk. Defending birds come near at a shorter distance (Curio and Onnebrink 1995) or remain near the nest (Amat and Masero 2004). However, if the birds respond to the predator's presence by an effort not to draw attention to the nest's proximity, they behave towards the nest predators more carefully, e.g. by substantially reducing the frequency of feeding (Ghalambor and Martin 2000).

A stronger reaction to specialized nest predators does not necessarily characterize all bird species. According to Ghalambor and Martin (2000), the intensity of the reaction also depends on their life history. In their experiments, it was only the shorter-lived white-breasted nuthatch (*Sitta carolinensis*) that responded to a nest predator (house wren—*Troglodytes aedon*) more vigorously, while the longer-lived red-breasted nuthatch (*Sitta canadensis*) acted cautiously in the presence of a predator of adult birds (sharp-shinned hawk—*Accipiter striatus*). The changes in the intensity of defence during the breeding season constitute an important proof of distinguishing nest predators from predators of adult birds as well. Nest predators endanger clutches from the very beginning, and the intensity of the defence against them remains the same throughout the breeding season (Green et al. 1990). On the contrary, predators of adult birds primarily imperil weaned chicks, and the intensity of the defence against them increases during the breeding season (Patterson et al. 1980; Green et al. 1990; Halupka and Halupka 1997).

Brood parasites represent a specific group of nest predators, differing from the others mainly by only being dangerous during the short early stage of breeding. This should be reflected in the corresponding timing of antipredator behaviour. What was primarily tested from the brood parasites was the recognition of the brown-headed cowbird (Nice and Ter Pelkwyk 1941; Robertson and Norman 1977; Smith et al. 1984; Folkers and Lowther 1985; Briskie and Sealy 1989; Hobson and Sealy 1989; Neudorf and Sealy 1992; Bazin and Sealy 1993; Mark and Stutchbury 1994; Gill and Sealy 1996, 2004; Gill et al. 1997a, b; Burhans 2001; D'Orazio and Neudorf 2008), as well as various species of the tropical cuckoos (*Urodynamis taitensis*, McLean 1987; *Surniculus lugubris*, Duckworth 1997; *Clamator glandarius*, Avilés and Parejo 2006; *Chrysococcyx lucidus*, Langmore et al. 2012). In all the aforementioned studies, the trial birds managed to recognize the brood parasite, adequately responding to it, which is not so surprising, considering the fact that the brood parasites were well distinguishable from the predators used or from harmless birds. The recognition of the parasitic European common cuckoo constitutes a more interesting issue, which has also been extensively tested. Although its mere recognition was also tested in many cases (Edwards et al. 1950; Moksnes et al. 1990; Bártol et al. 2002; Honza et al. 2006; Welbergen and Davies 2009), so was its distinguishing from a distinctly different predator (Campobello and Sealy 2010; Yang et al. 2014), considerable attention was paid on the other hand to the issue of its certainly nonaccidental similarity to the European sparrowhawk, which is the same for human as well as bird vision (Stoddard 2012). The similarity is so great that the cuckoo is often considered a mimic of the sparrowhawk (cuckoo-hawk mimicry hypothesis—Welbergen and Davies 2008). Despite the relatively small differences in appearance, the great reed warblers (*Acrocephalus arundinaceus*—Trnka and Prokop 2012) are capable of distinguishing the cuckoo and the sparrowhawk, so are the Eurasian reed warblers (Duckworth 1991; Welbergen and Davies 2008, 2011; Thorogood and Davies 2012), the red-backed shrikes (Ash 1970) and the willow warblers (*Phylloscopus trochilus*—Edwards et al. 1950). Studies testing birds which do not host the cuckoos bring an interesting comparison, as they ascertained that neither the titmice (Davies and Welbergen 2008) nor the sparrows (Trnka et al. 2015) were capable of distinguishing the grey morph cuckoo and the sparrowhawk, reacting to both with the same level of fear. The mentioned results therefore constitute a convincing piece of evidence that birds can learn to distinguish very similar enemies, if they have sufficient motivation to do so.

2.4   Different Species of the Same Ecological Guild

The number of studies dealing with the birds' ability to distinguish different types of predators of the same ecological guild has not been small (Table 1.3). However, they fail to have the same evidential value, since distinguishing the predators presented or researched in them is unequally demanding. Several studies compare responses to predators differing in sizes (Table 1.3). They usually pose a different level of threat for the potential prey as well. Each predator prefers a prey of an optimum size that can be effectively hunted and processed. Particularly in the case of bird predators, their own size determines an optimum size of the prey to a considerable extent. The possibility to separate the food niches in this way is even considered to cause significant sexual dimorphism in size (Overskaug et al. 2000), e.g. in some members of the Accipiter and Falco genera.

Nonetheless, what subsequently can be suggested is the possibility that the trial birds do not distinguish between individual kinds of predators, but only between their sizes. This particularly applies to all studies aimed at the variability of alarm calls, especially of the tits. It was already Apel (1985) who ascertained that presenting a mount of the sharp-shinned hawk to the black-capped chickadees results in a higher rate of calling than mounts of larger predators do. This finding was subsequently confirmed and further elaborated in several other studies on the captive black-capped chickadees (Templeton et al. 2005), as well as on the wild Carolina chickadees (Soard and Ritchison 2009) and the tufted titmice (Courter and Ritchison 2010). The chickadees and titmice responded to the presented predators of various size categories by uttering 'chick-a-dee' calls with different numbers and types of notes. Larger raptors less dangerous for small songbirds, such as the red-tailed hawk (*Buteo jamaicensis*), elicited calls with significantly more introductory 'chick' notes and fewer 'dee' notes, while smaller bird predators more dangerous for small birds (e.g. the eastern screech-owl) elicited calls with few or no 'chick' notes and significantly more 'dee' notes (Soard and Ritchison 2009; Courter and Ritchison 2010). However, they fail to bring clear evidence of different reactions to predators of the same size. Therefore, the alarm calls of the tufted titmice do not seem to provide any information directly on the kind of a predator discovered by them.

Some experiments focused on other responses than the vocal responses of birds appear to be similarly ambiguous. Curio et al. (1983) placed various live birds of prey to nest cavities of the great tits. Mobbing against individual predators differed in a minimum distance to which the tits dared to approach. The tits remained at the furthest distance from the sparrowhawk, whereas the closest they dared was to the tawny owl.

Ergo, they were at least able to distinguish a bird of prey from an owl, which is, on the other hand, easier than distinguishing between two types of raptors or the owls, since there is no need to make use of species-specific features, such as colouration. As a group, the owls differ from raptors in the overall shape of the body, the position of the eyes and the veil around them. The ability to distinguish between the birds of prey and the owls was also ascertained in other European tit—the willow tit (Kullberg 1998)—as well as in the American tufted titmouse (Sieving et al. 2010) or the Siberian jay (Griesser 2009).

Nonetheless, in his study Curio et al. (1983) also show that the great tits respond differently to two different owls as well. Even though they dared relatively close to the aforementioned tawny owl, they kept a greater distance from the pearl-spotted owlet (*Glaucidium perlatum*) considered the same as the Eurasian pygmy owl (*Glaucidium passerinum*). The tawny owl is not significantly larger than the sparrowhawk, while the owlet is the smallest Central European predator of birds; therefore its dangerousness could be underestimated. As opposed to the tawny owl, the Pygmy owl is a specialized predator of small birds though, which corresponds to the cautious behaviour of the tits. The ability to distinguish among raptors and the owls, differing from each other both in size as well as in their dangerousness for the species tested, was ascertained not only in various bird species but also in different contexts. Similarly to the aforementioned tits, the Siberian jays also reacted to the mounts of diversely dangerous birds of prey and owls placed at feeders (Griesser 2009). Several other studies confirm the same findings regarding various bird species in the field, whether using the presentation of mounts (Miller 1952; Altmann 1956; Nocera and Ratcliffe 2009) or only by means of observational studies (Buitron 1983; Winkler 1992). The ability to recognize predators similar in colour and differing essentially only in size was confirmed in experiments with birds reared in captivity (Palleroni et al. 2005). In the study, three trained raptors of different size categories (small-sized sharp-shinned hawk; medium-sized cooper's hawk, *Accipiter cooperii*; large-sized northern goshawk) flew over a paddock with freely moving fowls. Stronger reactions, such as sleeking and crouching, occurred most often in the presence of a large raptor, while displays of vigilance, like an erect posture with ruffling, constituted a more frequent response to a small bird of prey flying over. The fowls took up intermediate postures in the presence of a medium-sized raptor. Regarding the females, the intensity of their reaction was affected by the fact if they had chicks at the time of the experiment. Broody hens guarding the chicks were overall more aggressive, especially towards the small-sized sharp-shinned hawk.

Experiments with silhouettes of flying raptors provide very interesting results as well. Both captive domestic chickens (Evans et al. 1993b) and wild-caught birds, namely, the blue tits (Klump and Curio 1983) and the willow tits (Alatalo and Helle 1990), underwent these experiments. As opposed to experiments with mounted birds, or live birds mentioned hereinbefore, the trial birds responded to larger silhouettes more intensively than to smaller ones, specifically by reducing their motion and conversely by increasing the frequency or intensity of the alarm calls. The authors explain this by the fact that the size provided the trial birds with information how high the presented raptor was flying. A larger silhouette thus resulted in more fear since it presented a low-flying raptor, which was more dangerous for the potential prey. It was only the study of Alatalo and Helle (1990) that brought the opposite result. Here the tits called more intensively during a simulated flyover of a smaller silhouette. The authors explain this outcome by the fact that the alarm call is risky, and therefore the tits prefer being silent, only calling when the risk of drawing attention to themselves is lower.

Both interpretations look convincing, nonetheless leading to a question, why the tested birds responded to the differences in the actual size in all the experiments with mounts or live raptors described hereinbefore. Two answers can be suggested. In the experiments with mounts, the birds could distinguish not only the size of the predators but also the types. The second possible answer is more interesting. The silhouettes fail to provide the trial birds with any other feature than the size. If they were supplied, e.g. with eyes, a beak or claws, their similarly (un)easy distinguishability could provide information about them moving in similar height and varying in size. The study of Grubb (1977) brings such a conclusion, as he observed the reaction of the American coot (*Fulica americana*) to various naturally occurring raptors (*Buteo lineatus, Haliaeetus leucocephalus, Pandion haliaetus*). Even though the raptors flew at different distances and heights, the coots only feared the bald eagle (*Haliaeetus leucocephalus*), which is a well-known waterfowl predator. Other birds of prey (including airplanes) did not cause any fear. However, in the case of silhouettes not providing any detailed features, it cannot be easily distinguished whether they differ in the species, size or height of the flight. The potential prey thus only infers the height of the flight from the size of the silhouette, which is obviously the most reliable parameter for estimating the current dangerousness in the existing circumstances.

Even after we eliminate studies testing the differentiation of diversely sized predators, there still will be plenty of evidence regarding the ability to recognize individual species differing only in colour or in other features. Perhaps this was first accomplished in the work of Edwards et al. (1950), who found out that the willow warbler reacted to the mounts of two equally sized raptors (sparrowhawk, kestrel) of a different dangerousness located at its nest by a different intensity of mobbing. The ability to distinguish between these two birds of prey, nearly indifferent in size, but posing a different threat, was repeatedly ascertained in more recent studies as well. Both the shrikes (Strnad et al. 2012) and the tits (Tvardíková and Fuchs 2011, 2012) behaved much more cautiously towards the more dangerous sparrowhawk (a specialized predator of small birds) than towards the kestrel (catching mostly small mammals). Still the results of the experiments clearly show that the tits as well as the shrikes recognize predators in both species.

Observational studies provide further evidence. Three species of the lapwings observed by Walters (1990) mostly ignored the vultures and kites, responding to them only when they approached the chicks. They responded more intensively to the caracaras, falcons and hawks, which are their frequent predators. The behaviour of the southern lapwings (*Vanellus chilensis*) had an extraordinary evidential value, as they completely ignored the piscivorous specialist—the black-collared hawk (*Busarellus nigricollis*)—nonetheless reacting to other large hawks (*Buteo albicaudatus, B. magnirostris, Buteogallus urubitinga*) in the same way as to other birds of prey.

Chapter 3
Tools Used for Predator Recognition

The previous chapter cogently demonstrates that the ability to recognize predators is widespread among birds, involving not only distinguishing predators from the harmless animals but also mutually distinguishing individual groups and predator species from one another. This leads to a question what cues birds use for recognition.

3.1 Non-manipulative Studies

The studies testing the ability of birds to recognize predators discussed in the previous chapter can only help a little in finding the answer. Nonetheless, by the appearance of the tested predators and harmless animals used as a control, we can try to determine which of their differences (potential distinctive features) are available for recognizing, and therefore, we should aim our attention at them. Studies testing the distinguishing of representatives of various classes (Table 1.4), primarily ground and aerial predators, bring an insignificant benefit. Their difference is so substantial that anything can be used for recognition, from the body plan and covering of the body to various local features (muzzle vs. beak, etc.).

The owls and birds of prey rank among the most important and most frequently tested bird predators of adult birds. Both groups are relatively uniform in appearance, as they have similar body proportions (a fairly large head, a neck shorter than a tail and short legs) and are characterized by a few distinct local features (a shorter but thick hooked beak, strong legs with long claws). In addition, the owls have forward-facing large eyes with a veil, while birds of prey have a less distinct osseous superciliary ridge. All these features can be used to distinguish the representatives of both groups from harmless birds, which has been supported by a number of studies (Table 1.5). Nevertheless, this does not mean that birds tested in them could not have used other features typical for individual species in most of the studies, colouration in particular. This option is only excluded in species; the trial birds could

© The Author(s) 2019
R. Fuchs et al., *Predator Recognition in Birds*, SpringerBriefs in Animal Sciences,
https://doi.org/10.1007/978-3-030-12404-5_3

not have had prior experience with. However, none of these species have been used in any existing studies. Studies testing the recognition of the sparrowhawk and the grey morph of common cuckoo provide certain proof that birds indeed use the general features of raptors for recognition (Table 1.3). As far as their size, body proportions (a long tail) and colouration (wavy bars on the abdomen, a yellow eye) are concerned, both birds are almost identical. Where the trial birds did distinguish between them, they probably used the distinctly different shape of their beaks or a different length and claw strength to do so.

Nest predators have a lot less uniform appearance than birds of prey and the owls, even if we limit ourselves to the most frequently tested group—corvids. Leaving aside the more or less uniformly black-coloured species (most of the *Corvus* genus in Europe), we could only consider the strong straight beak as a mutual feature, besides the above-average size (within songbirds). Therefore, we would assume that distinguishing them (as a group) from harmless species (pigeons) constitutes a fairly difficult task. Yet, there is an indirect piece of evidence that birds are capable of it. Besides the Eurasian jay, the red-backed shrikes tested by Němec and Fuchs (2014) attacked another small-sized representative of the corvid family—the spotted nutcracker (*Nucifraga caryocatactes*). This species almost does not predate nests; therefore, it can hardly be assumed that half of the tested shrike couples attacking the nutcracker had had any negative experience with it. More likely, they evaluated the nutcracker as a potential danger, whereas its relatively massive beak could have been the clue to it.

Undoubtedly, size is the simplest feature singularizing different species of raptors and the owls, as most predators used when studying the information communicated in the alarm calls of the tits differed at that (Tables 1.1 and 1.2). Again, we cannot rule out that the birds tested managed to distinguish them on the basis of other features. This would only be the case if they responded identically to similarly sized species differing significantly in their dangerousness, for instance, due to their different food preferences. The representatives of the genus *Accipiter* in the experiments of Palleroni et al. (2005) differed in size as well. Using other features for distinguishing cannot be ruled out even in their case; however, at first glance it appears quite unlikely, since the colouration of all the used species is similar.

It was probably the size that played a decisive role in the behaviour of the shrikes towards potential brood predators of the *Corvus* genus, as they behaved towards them completely passively, as opposed towards the jay and nutcracker (Němec and Fuchs 2014). They differ from all three used members of the *Corvus* genus (*Corvus corone*, *Corvus frugilegus*, *Corvus corax*) in a significantly smaller size and thus a greater chance of an attack success. Additionally, the shrikes showed the greatest caution towards the largest one—common raven (*Corvus corax*). Interestingly, the other two *Corvus* species differ in the danger they pose to eggs or chicks. While the Carrion Crow (*Corvus corone*) is a major predator of nests (as well as the raven), the rook (*Corvus frugilegus*) seeks them only rarely. This would support the assumption that only size was decisive in the behaviour of the shrikes. Despite the uniform black colouration, an adult rook can be distinguished from the crow by a featherless base of the beak.

There are surprisingly few studies testing the ability of birds to distinguish between species of similar size, differing only in the colouration (Table 1.1). The distinguishing of the common kestrel and the Eurasian sparrowhawk is perhaps the most unambiguous case, having been ascertained in the tits in feeder experiments (Tvardíková and Fuchs 2011, 2012) as well as in the shrikes (Strnad et al. 2012) and also in willow warblers (Edwards et al. 1950) when defending their nests. These species have similar body build and size, bearing the typical common features of the vast majority of raptors (crooked beak, claws with long talons and an osseous ridge above the eyes). Therefore, it is evident that to distinguish them, the tested birds had to use their colouration. Nonetheless, the question is how. The kestrel and the sparrowhawk vary in a number of elements of colouration, by colour tones (e.g. brown vs. grey upperparts) on the one hand and by ornaments (e.g. lengthwise streaks vs. crosswise wavy bars on the chest and abdomen) on the other. Furthermore, their eyes also have a different colour (dun vs. yellow) and legs (pink vs. yellow). The individual elements are more (e.g. eye colour) or less (e.g. upperpart colour) distinct.

Birds can perceive the colouration as a whole, i.e. all the elements as more or less equal, or they can just focus on a single distinct element, unique for the particular predator. These options only represent the limit points of the gradient, and we can assume that the recognition will normally involve more colouration elements than one, albeit not all of them. Moreover, all the monitored elements need not be attributed an equal importance. What will probably play an important role will be their salience and uniqueness for that particular species of a predator. For instance, to recognize the sparrowhawk, its horizontal wavy bars on the chest and abdomen will undoubtedly be of a particular importance, as it is a distinct element (a contrast one, taking up a large area, clearly visible whether the bird is flying or sitting) as well as a unique element (apart from its relative—the northern goshawk—which is also a specialized predator of birds, albeit a much bigger one; the grey morph cuckoo and the barred warbler, *Sylvia nisoria*). However, the question is whether the birds would recognize the sparrowhawk if its horizontal wavy bars were removed or in reverse, if they would recognize the sparrowhawk in the kestrel—or in a completely harmless bird species (e.g. turtle dove)—if the horizontal wavy bars were added.

3.2 Manipulative Studies

To ascertain how tested birds use the range of available distinctive features, we must manipulate with individual colouration elements of the predators as well as with their combinations (including other features). The manipulation may involve removing the element (e.g. spots in the case of the kestrel; horizontal wavy bars in the case of the sparrowhawk) or replacing it with a feature of a harmless bird (e.g. the sparrowhawk's yellow eye with the red eye of the pigeon). Nonetheless, you can also manipulate with other features than colouration, for instance, a predatory beak, claws and size. Provided that the trial birds recognize the unmodified predator in such a prepared stimulus, i.e. they respond to it by a similarly intense antipredator behaviour (more details in Chap. 1), the manipulated feature is not necessary for recognizing the predator. Similarly, if the tested birds recognize a predator in the stimulus, i.e. if they respond to it by a more intense antipredator behaviour than to a harmless control (more details in Chap. 1), the manipulated feature is not necessary for recognizing the predator. In the event that the potential distinctive features are added to the decoy of a harmless bird (e.g. exchanging the pigeon's beak and eye for a raptorial beak and eye of the sparrowhawk or adding the sparrowhawk's horizontal wavy bars on the chest and abdomen of the turtle dove), we test whether the features in question are sufficient for recognizing a predator in general or a particular predatory species. Removing (replacing) features on a predator decoy and adding (replacing) features of a decoy of a harmless animal thus provide answers to different questions. Should the response to the predator decoy with removed (replaced) predatory (generic) features be stronger than the response to the decoy of a harmless animal with added (replaced) predatory (generic) characters, other features besides the tested feature (combinations of features) are evidently also used when recognizing a particular predator.

3.2.1 Silhouettes

Studying how birds recognize predators received much less attention than researching the results of this process, i.e. the ability to recognize predators and assess their dangerousness. However, an important group of studies originated in the pioneer period of ethology.

One of its founders, Konrad Lorenz, hypothesized that any specific animal behaviour is triggered by a specific stimulus, which he called the 'Auslöser' ('releaser' in English, Lorenz 1937a, b). The term 'sign stimulus' was derived from it by Tinbergen (1948), designating a signal (a body part or behaviour) emitted by one animal that causes a typical behaviour of another animal. Such signals subsequently control all the aspects of life from the epigamic behaviour (Lack 1943; Tinbergen 1948) to the nesting behaviour (Tinbergen 1951) to the antipredator behaviour and thus the predator recognition as well, which must precede the antipredator behaviour. Furthermore, the terms of 'key features' (Marr and Nishihara 1978) or 'salient features' (Schleidt et al. 2011) were also later used by authors concerned with recognizing predators as well as other animals (e.g. sexual partners).

Antipredator behaviour, which is spontaneous and not bound to a particular stage of life or season, hence offered one of the ways how to experimentally test the form and function of the sign stimuli. Their roles in activating antipredator behaviour (and recognizing predators) were first demonstrated in experiments with the silhouettes of raptors, where the reactions of the grouse (*Lagopus* and *Tetrao*), turkey (*Meleagris gallopavo*) and domestic chicken (*Gallus gallus* f. domestica) to the silhouettes of birds of different shapes were compared (Goethe 1937, 1940; Lorenz 1939; Krätzig 1940).

The results showed that it was the relative length of the silhouette's neck that played a decisive role in triggering antipredator behaviour. When the silhouette was moved in such a direction that it created the impression of a short neck in the front and a long tail at the back (predator), it gave rise to stampede behaviour in the tested poultry. When the same decoy was moved in the opposite direction, therefore appearing like a bird with a long neck and a short tail (goose), no antipredatory behaviour emerged. Other potential signs or the wing and body shapes of the silhouettes failed to affect the responses of the trial birds.

However, in the review of Lorenz and Tinbergen's experiments, Schleidt et al. (2011) observed that Tinbergen and Lorenz had elaborated more studies on the 'short neck', nonetheless had drawn different pictures in them not consistent with each other. Ergo, it is not entirely clear what was used and when. Moreover, according to the studies, Lorenz himself did not so much consider the 'short neck' as the antipredator response 'releaser' but rather a 'slow relative velocity of flight'. The authors of this review (Schleidt et al. 2011) themselves came to the conclusion that captive bred turkey chicks without any experience from the wild actually react rather to the silhouettes resembling predators while feral turkeys with individual experience with predators are virtually afraid of any object flying over them.

Despite this being a very traditional topic, the question of whether birds are capable of distinguishing predators from harmless birds by silhouettes cannot be considered solved, though, undoubtedly, such a capability would be very useful during real encounters with predators, since it could accelerate the appropriate response. On the one hand, it would increase the chance of escaping, and on the other, it would reduce unnecessary time and energy losses. Further experiments could thus bring some interesting results. They should be focussed on increasing the credibility of the predator decoys and their movement. However, the experiments should primarily use birds captured in the wild. While there have been few exceptions, the existing studies have worked with individuals coming from farms, and, in addition to that, these were mainly members of domestic poultry.

3.2.2 Particular Features: Colouration

In the subsequent decades, studies focussed on finding and checking the sign stimuli used to recognize predators were not very frequent, probably the most extensive research on this subject being carried out by Curio (1975), who observed the intensity of alarm calls (call rate) of the European pied flycatcher (*Ficedula hypoleuca*) against the modified decoy of the red-backed shrike. Though this species of a songbird feeds mainly on large insects, it is capable of catching even small vertebrates. As part of his study, Curio tested a lot of individual features, particular elements of the colouration being the most frequent. At first, he presented decoys of the shrike male with some colour modification of the black stripe through the eye, which is probably this predator's most distinct colour element. Naturally, an unmodified control mount of the shrike aroused the strongest alarm response among the flycatchers (Fig. 3.1). The reaction to a mount with a red stripe through

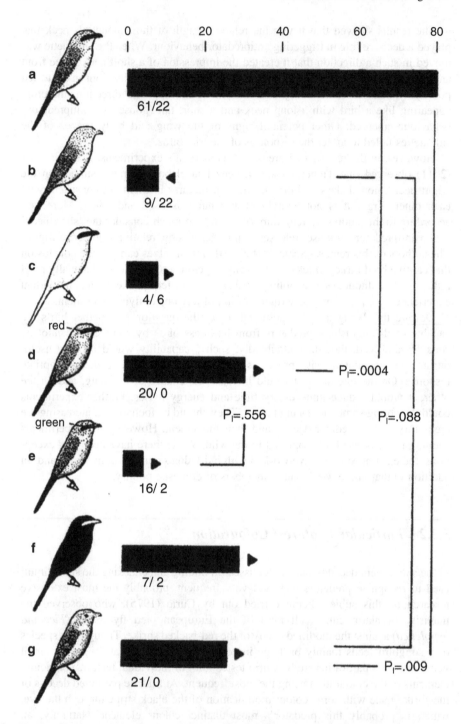

Fig. 3.1 The number of calls per minute (black bar) of pied flycatcher (*Ficedula hypoleuca*) to the shrike dummy with manipulated presence, colouration and contrast of the eye stripe. Below each bar, the number of experiments conducted in first/second half of the nestling period is mentioned. *P* values from two-tailed M-test. (Curio 1975 with permission)

the eye did not significantly differ from the reaction to the control, while a mount with a green stripe caused a significant decrease in the reaction intensity. Moreover, the effect of the contrast between the stripe through the eye and the rest of the head was tested. Detracting the saturation of the stripe failed to produce major changes in the flycatchers' response. Only when the stripe completely merged with the ground, the flycatchers stopped emitting alarm calls.

These results suggest that it is the presence of a distinct colour feature rather than its style that is of a key importance. Nevertheless, it must not probably be too far away from the actual appearance of the shrike as such. Red or reddish-brown colour along with various shades of grey is included in its colouration, while any shades of green are not; therefore a stripe through the eye in this colour is unlikely to cause recognition of the shrike. However, the mere presence of an unmodified feature is not sufficient either, since a white rectangular bar with a stylized eye stripe—but lacking a bird body shape—failed to elicit any alarm responses from the flycatchers. While the presence of a credibly coloured stripe through the eye is essential for recognizing the shrike, it is not a 'releaser' as perceived by Lorenz (1937a, b) since a releaser should trigger antipredator behaviour under any circumstances.

The position of the eye stripe is important as well. The flycatchers responded to the decoy with the eye stripe shifted from the forehead to the nape and even quite stronger than to an unmodified control decoy. Other modifications (stripe below the neck, on the abdomen or vertically on the nape) elicited only slight alarm responses. Presentations of the male shrike decoys with a colour-wise radically modified bodies produced similar results as well. The flycatchers only reacted strongly to an unmodified control decoy. When the shrike was removed of all the colouration elements and the black stripe through the eye was left on a uniform white body, the intensity of the flycatchers' alarm calls dropped dramatically. A uniformly white decoy without an eye strip raised no reactions.

These results again show that the eye stripe itself is not sufficient for recognition. It must be placed on the correct spot of the body, supplemented by other elements of colouration. On the other hand, the eye stripe is necessary for recognition. If the flycatcher encountered the shrike without the stripe through the eye, it would probably fail to appraise the shrike as a threat. This is particularly interesting as the shrike actually provides a feature directly referring to its potential danger-ousness—in the nomenclature of behavioural ecology, we could call it an honest feature. It is its big and sturdy beak hooked on the end, which could play the same role in recognition as the hooked beak of the owls and raptors (see Chap. 2). However, all European shrikes (seven species) have the stripe through the eye, and consequently there is no reason to use less distinct, albeit 'honest', features for recognition.

As mentioned hereinbefore, dark horizontal wavy bars on the otherwise pale underparts is a characteristic colour feature of the most dangerous European predator of small birds—the sparrowhawk. Therefore, several authors tried to test its impor-tance for recognizing this raptor by a potential prey. In the work of Veselý et al. (2016), the tits on a feeder were presented with the decoys of the European sparrowhawk with completely or partially modified colouration. A decoy with the characteristic horizontal wavy bars on the abdomen removed, as well as a decoy coloured as the harmless great tit and European robin, caused the same fear in the tits as the unmodified sparrowhawk. Only a decoy with an artificial purple and white checkerboard pattern was not perceived as a greater threat than a harmless control (pigeon). Therefore, this study failed to prove the necessity of the abdominal horizontal wavy bars for recognizing the sparrowhawk. The same results were reached by Welbergen and Davies (2008), who also presented a decoy of the sparrowhawk on a feeder deprived of the abdominal horizontal wavy bars. As in the previous study, the tits failed to fear it less than the unmodified sparrowhawk.

The results of experiments with the sparrowhawk are therefore fundamentally different from those obtained in experiments with the red-backed shrike. While the horizontal wavy bars on the underparts are not essential for recognizing the sparrowhawk, the stripe through the eye is important for recognizing the shrike. Several explanations are possible. First of all, the decoy with removed horizontal wavy bars still has at least one feature characteristic for the sparrowhawk—the yellow eye. No other European raptor of a comparable size is endowed with it. Nonetheless, at first glance it would seem unlikely that such a detail could suffice for the recognition of the sparrowhawk (however, see Chap. 2). Still, it cannot be ruled out that the tits did not recognize the sparrowhawk with removed horizontal wavy bars in the decoy, but a predator of a species unknown to them, to which they reacted equally cautiously as to the sparrowhawk. The modified decoy offered a variety of features, ranging from the body proportions to the length of the claws, common to all European birds of prey. Cautious behaviour towards an unknown predator also appears likely, since birds visiting a feeder are generally quite cautious, with any suspicious object causing fear in them. The limits of experiments with untrained birds can be seen here. The distinguishing of any 'objects' can be ascertained in them only if they give rise to a different behaviour (see Chap. 1). It would be interesting if the tits were offered a direct choice between a modified and unmodified decoy in a two-feeder experiment. Moreover, it would be worth to observe their habituation, which should be faster in the case of an 'unknown predator'.

The extraordinary colouration resemblance of the grey morph cuckoo brings an interesting opportunity to study the importance of the horizontally barred underparts for recognizing the sparrowhawk (see Chap. 2). The cuckoos generally tend to adapt their colouration to the sympatrically occurring species of raptors (Gluckman and Mundy 2013; Thorogood and Davies 2013). Besides, some observational data can be used as well. Lyon and Gilbert (2013) observed mobbing the cuckoos by a species the cuckoo almost does not parasite on (barn swallow, *Hirundo rustica*) or does not know (American bushtit, *Psaltriparus minimus*). According to the authors, in both cases they confused it with the European sparrowhawk or in the case of the American bushtit with some of the local representatives of the *Accipiter* genus (e.g. *Accipiter striatus*). According to Langmore et al. (2012), the tits (great tit and blue tit) also react to the cuckoo similarly cautiously like they do to the sparrowhawk, yet even these songbirds are not parasitized on by the cuckoo. On top of that, Liang and Møller (2015) ascertained that the barn swallows defend the nest against the cuckoo and sparrowhawk more intensively in Denmark than they do in China. The authors explain these results by a fact that sympatrically more species of brood parasites occur in China, and therefore they are harder to recognize for potential hosts. Nonetheless, the sparrowhawk being more abundant in Denmark than it is in China may also be the reason, as it constitutes a better-known danger for the swallows. Seemingly, while the horizontal wavy bars on the upperparts are not necessary for recognizing the sparrowhawk, together with the yellow eye, it is sufficient for the cuckoo to be recognized in a harmless (for the tested species) cuckoo, despite the latter differing in a completely dissimilar shape of the beak and claws at least.

However, the Eurasian reed warbler, commonly parasitized on by the cuckoo, responded in the vicinity of the nest to a cuckoo mount by much stronger alarm calls than to a European sparrowhawk mount and a harmless species control (Eurasian teal, *Anas crecca*), which indicates that the reed warbler reliably distinguishes between them (Welbergen and Davies 2008). Several other studies testing the reactions of potential cuckoo hosts drew similar conclusions (Edwards et al. 1950; Ash 1970; Duckworth 1991; Thorogood and Davies 2012). In addition to that, Trnka et al. (2012) ascertained that the horizontal wavy bars on the cuckoo's abdomen are not necessary for recognizing it. The great reed warblers attacked a cuckoo mount with the horizontal wavy bars only slightly less than an unmodified cuckoo mount, suggesting that they recognized it owing to another feature. Therefore, if the birds are motivated to distinguish between the cuckoo and the sparrowhawk, they can acquire this ability. The question remains whether they learn to distinguish by observing experienced individuals or whether it is inherited to some extent.

However, the previous conclusion can be cast doubt upon by the results of other experiments demonstrating that the similarity of the grey morph cuckoo with the European sparrowhawk really functions as a means of protection against attacks by the host birds. The reed warblers dare to draw closer to the cuckoo without the horizontal wavy bars on its underparts than to the unmodified cuckoo (Welbergen and Davies 2011). On the contrary, when a control collared dove mount decoy (*Streptopelia decaocto*) had the horizontal wavy bars added, the warblers approached it less than an unmodified decoy (Welbergen and Davies 2011). This variance with the foregoing experiments could be explained by a different ability of individual birds to distinguish the cuckoo and the sparrowhawk, which would thus support its acquisition by learning.

Nonetheless, the effect of the horizontal wavy bars is not universal in the reed warblers either, as they approached a sparrowhawk mount with the wavy bars on the underparts removed with the same caution as an unmodified sparrowhawk (Welbergen and Davies 2011). This result corresponds with the above-described studies on the tits, and the explanation will probably be similar. The reed warblers either recognize the sparrowhawk by other features, which the authors tend to think, or they fail to recognize the sparrowhawk but are cautious towards the decoy as it bears all general raptorial features, thus can be assessed as an unknown and potentially dangerous raptorial species.

The conclusions obtained in experiments with the reed warbler do not have to apply to other hosts of the cuckoo. Trnka and Prokop (2012) found out that the great reed warblers did attack the cuckoo more intensely than the sparrowhawk, yet only when the *decoys* of both intruders were presented concurrently. When presented separately, the intensity of the attacks was not significantly different owing to more frequent attacks on the sparrowhawk. The authors explain that the great reed warblers are extremely aggressive due to their physical disposition, generalizing the cuckoo's dangerousness to all intruders in defending the nest. In this case, however, the intensity of the attacks on the cuckoo and sparrowhawk should not differ even when presented concurrently. If the great reed warblers only attack intensively a separately presented sparrowhawk, a more logical explanation is that they do not generalize the dangerousness of the cuckoo but its appearance, i.e. they recognize the sparrowhawk as the cuckoo. Nonetheless, the cause for their decreased ability to distinguish the grey morph cuckoo from the sparrowhawk can lie in their better ability to drive any intruder away from the nest. Trnka and Grim (2013) also tested the response of the great reed warbler to the brown morph cuckoo. They attacked the cuckoo a lot more intensely than the harmless collared doves, thus clearly recognizing a danger in it. However, the question is what kind of danger they recognized, since the intensity of the attacks was lower than attacks on the grey morph cuckoo, not significantly differing from the intensity of the attacks on the sparrowhawk and kestrel. Obtaining a more accurate response would require a manipulation with the features of the brown morph cuckoo (colouration) as well as the kestrel (commonly occurring features in raptors).

To sum up based on the results of the existing experiments, birds undoubtedly use distinct particular elements of colouration for recognizing predators (and brood parasites). However, they may not be necessary or sufficient. If they are missing, they may be replaced by other less distinct features—at least at first glance. On the contrary, if such 'complementary' features significantly differ, the distinct colouration elements may not work per se. Evidently, the result depends on the 'reliability' of particular elements. If it is necessary to distinguish a predator from a brood parasite, they may be (to some extent or by some individuals) ignored, as it happens in the case of the horizontal wavy bars on the underparts of the sparrowhawk. On the other hand, if they are present at all the representatives of a group posing similar dangers, such as a stripe through the eye of the European shrike or a typical beak and claws of raptors, they can be relied upon. Naturally, the shrike must not be green, and raptor must not be evenly checked. When recognizing predators, untrained birds thus behave similarly to the pigeons trained to recognize a human figure in experiments of Aust and Huber (2001, 2002). Among other things, clothes constituted an important key feature for them; nonetheless, they had to be placed on an object more or less similar to a human figure.

3.2.3 Particular Features: Size

The importance of the physical size for recognizing different species of predators suggested by non-manipulative studies (see Sect. 3.1) should be supported or disproved by purposeful manipulation as well. Evans et al. (1993b) confronted inexperienced chicks with differently sized raptorial silhouettes projected on a monitor on the ceiling of their sleeping quarters. Large silhouettes evoked the most intense alarm calls; other behaviour also changed according to the size of the projected silhouettes, though. When a small silhouette was being projected, only the eye fixation of the chickens on the predator and turning heads contemplating the silhouettes could be observed. The silhouettes of a medium size caused a slight crouch, whereas the projection of the largest predators resulted in such intensive crouching of the chickens that their bodies literally touched the ground. The authors' interpretation of this result is that the chickens perceived the size as an indicator of the height, thus responding more intensively to the nearer predator.

Variously large decoys of the northern goshawk were used by Klump and Curio (1983) for testing wild-caught blue tits. The decoys were moved on a rope at a height of 4 m. The tits responded to the goshawk of a usual size by loud alarm calls and a movement inhibition lasting up to 3 min. A smaller decoy elicited a soft and brief alarm call 'seet', causing a movement inhibition for less than 1 min. Similar to the previous study, the authors suggest that the tits perceived differently sized decoys more like predators at different distances, adapting their antipredatory response to that. Nonetheless, the opposite interpretation is possible as well—i.e. the tits recognized the right size of the decoys actually responding more intensively to a small predator rather than a large one. This would be consistent with using the 'seet' alarm call for a remote (i.e. small) silhouette, which the tits use as a signal of an imminent danger. In such a case they would more likely recognize the sparrowhawk in a small goshawk, as they do not differ in their colouration so much, and the tits are the sparrowhawk's typical prey. Similarly, we could polemicize the conclusions of the previous study, since larger raptors (e.g. goshawk) contrarily represent a greater risk for the domestic fowl.

Nevertheless, the above-mentioned and other similarly formulated studies cannot clearly decide on the relevance of the size for distinguishing various species. When presenting flying decoys, we can never unequivocally decide whether the size of the decoys refers to a body size or distance. The experiment would have to be contrived in such a way that the tested bird would find itself in a clearly defined distance from the predator, therefore by using a sitting decoy. It was only Curio who worked with such decoys (1975). In his experiments, a smaller red-backed shrike decoy with an unchanged colouration elicited a similar reaction from the European pied flycatcher as an unmodified decoy.

3.2.4 Particular Features: Eyes and Beaks

Many species of the owls and raptors have very distinct eyes, contrasting with the head colouration. Additionally, a distinctly coloured bare skin often occurs around the eye, underlining its markedness even more.

It has also been repeatedly established that birds are capable of identifying whether a predator is watching them and therefore, whether it poses an imminent danger for them. Watve et al. (2002) ascertained that it is important for the green bee-eater (*Merops orientalis*) what angle the human eye look forms with the straight line leading to its nest, whereas the angle formed between the body and the nest is of no significance. Similar findings were obtained by Carter et al. (2008), when they tested the effect of a direct human look on feeding common starlings (*Sturnus vulgaris*). When the look of a 'human predator' was averted, the starlings returned to foraging, subsequently consuming more of it in less time. Therefore, it is not surprising that the presence and colour of the eyes are often considered sign stimuli allowing the recognition of various species of predators, and their importance has been experimentally tested as well. However, most of these studies are of an older date.

Nice and Ter Pelkwyk (1941) tested the response of the song sparrow to variously modified cardboard models of the barred owl (*Strix varia*) concluding that the head is the most important for recognizing the model as the owl; nevertheless the mere absence of the eyes fails to decrease the fear of the model. Conversely, Smith and Graves (1978) showed that the absence of the head—or only the eyes on the Eurasian eagle-owl (*Bubo bubo*) mount—greatly reduces the intensity of mobbing by the barn swallows. In experiments with inexperienced chickens, Scaife (1976) ascertained that a kestrel mount deprived of eyes as well as the brown kiwi (*Apteryx australis*) mount with the kestrel's yellow eyes added causes antipredator behaviour (retreat into hiding). However, both these mounts elicited a weaker response than an unmodified kestrel. Besides the colouration, Curio (1975) also dealt with the importance of the eyes in his comprehensive study. A mount of the shrike without the eyes aroused no antipredatory reactions in the flycatchers. In experiments with the pygmy owl, the flycatchers virtually ignored the mount, if one eye was covered with feathers. However, when both eyes were covered, the flycatchers began to respond, even though the increase in their activity was not significant and could not match the response to an unmodified pygmy owl. The author saw this increase in the restoration of the mount's symmetry. When the eyes of the pygmy owl were transferred on a shrike mount, they increased the intensity of the flycatchers' antipredator behaviour compared to their response to an unmodified shrike. However, when both eyes were on the same side of the head, the intensity of the flycatchers' response decreased. Finally, the flycatchers reacted to a pygmy owl mount with triangular eyes almost as intensively as to an unmodified control decoy.

Some very interesting results were provided by the testing of the significance of the yellow eye for recognizing the cuckoo by the great reed warbler (Trnka et al. 2012). The cuckoo mount with a black eye instead of a yellow one was practically not attacked, although its colouration was not modified. This result is surprising mainly due to the fact that the horizontal wavy bars on the underparts, which constitute at least an equally distinct feature, were not necessary for the recognition (see Sect. 3.2.2). The authors explain this by saying that the potential host mostly watches the brood parasite from above, therefore not seeing its abdomen. The yellow eye is thus a safer feature for the recognition in this case.

A hooked beak is another characteristic feature of raptors (and the owls), allowing them to carve the prey. Therefore, it can be designated as an 'honest signal' in behavioural ecology terminology, as no representatives of both groups lack it, with the odd exception. As opposed to the eyes, the shape and colour of the beak of the owls and raptors are fairly uniform. Ergo we can assume that they can be used more in recognizing the affiliation with both groups of aerial predators than distinguishing different species. The studies testing the significance are older as well, and their conclusions must be accepted with some caution.

In the experiments of Curio (1975), the extension of the beak failed to change the flycatchers' response to the red-backed shrike. Nonetheless, in the case of the pygmy owl, the extension of its beak resulted in a partial reduction of antipredator behaviour intensity. Smith and Graves (1978) ascertained that a mount of the great horned owl without a beak elicits much weaker mobbing in the swallows than a mount with a beak. Gill et al. (1997a) tested the key importance of the beak in recognizing the brood parasite brown-headed cowbird. A beak of the starling was attached to its mount. Like the beak of the cowbird, the beak is dark; nonetheless it is longer and thinner. The yellow warblers (*Dendroica petechia*) responded a lot more intensively to the unmodified mount with the original beak.

Beránková et al. (2014) presented the tits in laboratory conditions with the decoys of the pigeon and sparrowhawk with mutually exchanged eyes and beaks. The presence of the pigeon's eyes on the sparrowhawk reduced the level of fear it aroused. However, the presence of the sparrowhawk's eyes did not increase the level of fear the pigeon caused. On the contrary, the presence of the pigeon's beak did not decrease the fear of the sparrowhawk, but the presence of the sparrowhawk's beak increased the fear of the pigeon. This result reflects the fact that the beak serves as a particular raptorial feature for the tits, while the yellow eye serves as species-specific feature of the sparrowhawk. Therefore, the yellow eye on the pigeon fails to increase its 'dangerousness' (it has no other features of the sparrowhawk), while the absence of the eye decreases the dangerousness of the sparrowhawk (it lacks an important species-specific feature). Similarly, the sparrowhawk's beak increases the 'dangerousness' of the pigeon (none of the other features it has distinctly contradict with the affiliation to birds of prey), while the absence of the beak fails to decrease the dangerousness of the sparrowhawk (it has all the species-specific features).

A different approach was opted for by an earlier study of Edwards et al. (1950) dealing with the recognition of the cuckoo, with the whole head manipulated. The results are surprising in retrospect. The tested willow warblers did not attack a cuckoo torso deprived of its head. On the other hand, the separate head itself elicited an intensive mobbing reaction. This result is in accord with pioneer studies (e.g. Lack 1943) aimed at recognizing sexual partners and rivals. Nonetheless it seems to be in discord with newer studies, which indicate that key features only work if they occur in the context of other features. It is quite surprising that these experiments have not found a successor yet, since the reliable testing of the effect of isolated key features undoubtedly constitutes a necessary step in understanding the cognitive processes in untrained animals.

Other studies produce contradictory results as well; the causes can only be speculated about. We may find them in the birds tested (experienced vs. inexperienced), various dangerousness of the predators used (e.g. barred owl vs. Eurasian eagle-owl) and the monitored parameters of their behaviour but also in the different meaning of the tested features in various predators, which is indicated by the results of Beránková et al. (2014). In this respect, the results of Gill et al. (1997a, b) are interesting as well. The crucial importance of the beak for recognizing the brown-headed cowbird may lie in the fact that this uniformly coloured brood parasite does not offer any other features to distinguish it from medium-sized songbirds. This stands out in comparison with another brood parasite—the common cuckoo—discussed in the previous chapters. In any case, the importance of the eyes and a beak for recognizing groups and species of predators would be worthy of further attention.

3.2.5 Particular Features: Multiple Features

Studies discussed in the previous chapters show that the effect of a single, albeit a very distinct, feature usually depends on the context of other features. The horizontal wavy bars on the underparts of the sparrowhawk and the cuckoo provide perhaps the nicest sample owing to the considerable attention paid to it. If we want to learn about the meaning of the context more precisely, we need to manipulate more features at the same time, which will allow testing their various combinations purposely and systematically.

Beránková et al. (2015) presented the tits with modified sparrowhawk mounts differing in the overall colouration and size. Two sets were tested—one in the sparrowhawk's natural size and one scaled to the size of the great tit. The mounts were presented as sitting predators; therefore the possibility that the tits would perceive them as a differently distant danger could be ruled out. The colouration was (1) natural to the sparrowhawk, (2) natural to the pigeon (a harmless bird of the sparrowhawk's size, (3) natural to the robin (a harmless bird of the great tit size), and (4) natural to the great tit (Fig. 3.2). The tits exhibited fear in the presence of both decoys coloured as the sparrowhawk. When the colouration was maintained, the size played no role. Similarly, both predators coloured like the great tit caused no fear. Here the conspecific colouration eliminated the presence of general predatory features (beak, claws) as well as the presence of a specifically coloured sparrowhawk eye. The effect of the size had an impact only in case of decoys coloured as the European robin and the pigeon, where the decoys in an unnatural size (a large European robin, a little pigeon) aroused more fear than the naturally sized decoys.

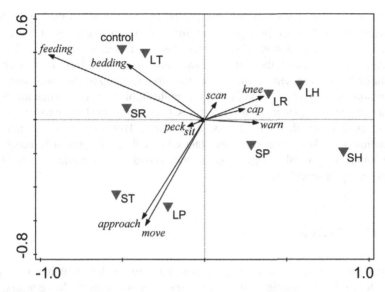

Fig. 3.2 Position of particular dummies on the first and second canonical axis of the principal component analysis created on the basis of particular behavioural responses of the great tits performed towards these dummies. Control = no dummy; *ST* small tit, *SR* small robin, *LT* large tit, *LP* large pigeon, *SP* small pigeon, *LR* large robin, *LH* large hawk, *SH* small hawk (Beránková et al. 2015 with permission)

The size thus played a role only in the case of decoys coloured as the birds the tits can encounter but which are not interesting for them under normal circumstances. When these species were presented in the 'wrong' size, the maintained raptorial or sparrowhawk's features attracted the attention of the tits.

Němec et al. (unpublished) simultaneously tested the significance of potential general key features of raptors (a hooked beak, long claws and an eye with an osseous eye ridge) and a species-specific plumage colouration for the recognition of the kestrel by the shrikes defending their nest. The key raptorial features remained raptorial in one set of decoys (made of plush), while in the case of the second set, the key raptorial features were replaced with those of the pigeon (an eye without the supraorbital ridge, thinner straight beak and pink legs with short straight claws). Each set contained three different modifications of the decoys: (1) a commonly coloured one, (2) a simplified one (devoid of the black spotting and other black elements in the colouration) and (3) a completely different one (a distinct—and quite exotic for central Europe—colouration of the South Asian raptor black baza (*Aviceda leuphotes*)) were used. Stuffed kestrels and pigeons served as the control (to verify the credibility of the plush decoys). In the case where the key raptorial features (the original kestrel beak, eye and claws) were kept on a normally coloured plush kestrel, this decoy was considered the kestrel by the shrikes, and they attacked with the same intensity as the stuffed kestrel. They attacked the simplified decoy with a slightly lower intensity, while they attacked the decoy in the black baza colour with the key raptorial features only minimally. In contrast, when the key features were replaced by those of the pigeon, the shrikes failed to attack any of the decoys regardless of its colouration. Even in the case of a decoy with an unchanged colouration of the kestrel, the shrikes showed no interest in it, feeding their chicks undisturbedly in its presence. The shrikes therefore recognized the species-specific colouration, even in a quite modified form, but this took place only when the decoy concurrently had the general raptorial features. Conversely, the key raptorial features themselves did not elicit an aggressive response. Therefore, it seems that the recognition of the kestrel occurs in two stages. In the first one, a raptor is recognized based upon the general features, and in the second one, a particular species is recognized upon specific features.

3.3   Conclusions

At first sight, the reader may view the results of the studies dealt with in this chapter as a patchwork, impossible to draw a general conclusion from. Nonetheless, we should attempt to make some kind of a conclusion. Evidently, birds do use distinct particular features for recognizing predators. They can even be essential for the recognition yet usually insufficient. Therefore, this is not a matter of the 'Auslöser', as defined by Lorenz (1937a, b). They do not have the same absolute effect, as, for instance, a tuft of red feathers had in the experiments carried out by with intraspecies recognition in the case of the European robin (*Erithacus rubecula*) (Lack 1943).

However, it seems that more often such a feature is rather necessary for the recognition than sufficient. In other words, it is more frequent that a predator without a distinct feature is not recognized as a predator than a harmless species with a distinct feature is. In a superficial reflection, this difference is somewhat surprising. If the distinct feature is not sufficient for the recognition, it should not even be essential. However, a hypothesis how the recognition process corresponding with the ascertained differences could proceed can be suggested. In the case of a raptor (e.g. the sparrowhawk) without a distinct species-specific feature, the hypothesis might look as follows, 'This bird is a predator, it looks like the sparrowhawk, but it has no horizontal wavy bars on its abdomen, ergo it is not the sparrowhawk'. In the case of a harmless bird species (e.g. turtle dove) with a species-specific distinct raptorial feature (e.g. of the sparrowhawk), the assessment would be, 'This is not a bird of prey, therefore it is not necessary to pay attention to its colouration'. Such a process would explain the trials of Němec et al. (unpublished), in which the shrikes failed to respond to the kestrel with the pigeon eye, beak and talons. Therefore, they paid no attention to its species-specific colouration, in spite of undoubtedly knowing it.

The existing results also show that the effect of distinct features is influenced by the degree of their specificity. General features (such as common raptorial features) have a bigger effect than special features (such as the species-specific colouration of the kestrel). The pigeon with a raptorial beak aroused a certain degree of fear (of course a lower degree than an unmodified sparrowhawk), while the kestrel without a raptorial beak and talons did not. That makes sense. All birds probably encounter previously unknown bird species during their life. The common raptorial features constitute a useful clue for their classification in the 'bird of prey' category. Colouration, which is very species-variable, is not such a feature—not even in the case when it resembles the colouration of a raptor or, contrarily, the colouration of a harmless bird.

Chapter 4
General Principles of the Objects Recognition

As evident from the previous chapter, behavioural ecology studied the predator recognition process rather subjectively. It followed up on the work of the founders of the European ethological school, mainly focusing on the role of individual distinct features. Cognitive ethology deals with studying recognition processes in a much deeper way. However, it is not about recognizing predators or other relevant objects by untrained birds. Cognitive ethology followed up on the founding work of the American ethological school addressing the learning process (Watson 1913; Skinner 1931), which developed a precise methodology based on conditioning learning. In the case of studying recognition processes, the researched animals are in the simplest option trained to distinguish between two groups of objects, so that they are rewarded for the answer to one of them (positive stimuli) while not being rewarded for the answer to the other (negative stimuli). After having reached a certain level (proportion) of correct responses, the experimental animals are presented with new groups of objects that are modified in one or more features to determine whether they categorize these objects into groups learnt during the training.

As far as birds are concerned, domestic pigeons (*Columba livia* f. domestica) are most commonly used in these experiments. The experiments are usually carried out in the so-called Skinner box, which provides maximum standardized conditions. The animals tested here are not subjected to any other stimuli than the tested ones in a well-defined procedure. Their nature can be much varied; nonetheless, the vast majority of studies involve a two-dimensional projection of two- and three-dimensional 'objects' displayed on a computer monitor. More or less schematized drawings and photographs of simple geometric shapes (e.g. Biederman 1987 or Kirkpatrick-Steger et al. 1996) and various human creations of living organisms (e.g. Lubow 1974 or Lazareva et al. 2004) as well as persons (e.g. Herrnstein and Loveland 1964 or Aust and Huber 2002) were used.

Behavioural psychology (Shettleworth 2010) assumes that two cognitive abilities—discrimination and categorization, are simultaneously applied in any recognition process, whereas discrimination means the distinguishing of the object to be recognized from objects with different properties and categorization means its

R. Fuchs et al., *Predator Recognition in Birds*, SpringerBriefs in Animal Sciences,
https://doi.org/10.1007/978-3-030-12404-5_4

classification into a group of objects with the same properties. In the case of discrimination, we are interested in what cues the animal uses in order to distinguish the objects; in the case of categorization, we want to know how the animal generalizes new objects onto objects already known.

Even though both abilities in the recognition process are applied simultaneously, we will first deal with discrimination, as most of the studies mentioned in the preceding chapters are devoted to dealing with the search for features that birds use to distinguish predators from harmless animals.

4.1 Particulate Feature Theory × Recognition by Components

The beginnings of systematic research of cues used by birds to discriminate objects can be put down to the 1980s. At that time, two theories were developed describing how birds work with potential cues available to them for distinguishing objects. The older one of the theories—the Particulate Feature Theory (PFT) assumes that birds (and other animals) perceive objects as a disarranged set of individual local features (Cerella 1980). A mere presence of these features determines the distinguishing of a particular object from others (e.g. a predator from non-predators). A more recent theory called Recognition by Components (RBC) contrarily assumes that objects are not defined only by the presence of local features, but also by their relative location (Biederman 1987). At first glance, both theories contradict each other; however, the results of experimental studies show that the contradiction is only seeming.

Cerella based his theory on the results of his experiments with pigeons (Cerella 1986). He trained them to distinguish the images of Charlie Brown from other heroes from the Peanuts comic book. In the experiment itself, all the characters were horizontally split into three equal parts that were mutually mixed (Fig. 4.1). In the experiment the pigeons distinguished Charlie Brown's character from the other

Fig. 4.1 Intact, scrambled and incomplete test figure of Charlie Brown (Cerella 1986 with permission)

characters with almost the same success they achieved in training with unmodified drawings. Therefore, the mixing of body parts failed to affect their ability to discriminate and categorize. Based on these results, Cerella (1986) concluded that the presence of individual local features is sufficient for the recognition of the object and their relative location is of no consequence.

Biederman (1987) formed the RBC theory on the basis of studying human recognition processes. According to him, we can describe any unknown object in detail by mentally taking it apart into parts of known shapes, such as a cylinder or a cube. The RBC theory calls these and similar geometric shapes geons. There is a limited amount of geons and any object can be put together from them, regardless in which angle we view the object. (Biederman 1987). However, as opposed to the PFT theory it depends on their relative location. This is due to the fact that while features considered by the PFT theory are (if possible) unique, geons represent universal basic shapes. In addition, Biederman (1987) proposed a sequence of steps for the RBC theory, which should be necessary for recognising any object. First, we notice its outline, then colours, structure and brightness. Nonetheless, at the same time we also perceive possible symmetry and, above all, differentiate basic shapes (geons) that the object is comprised of. We further analyze the information we collect, primarily concentrating on the concave areas, and in particular on those that show essential properties. It is exactly these areas that are instrumental in discriminating between objects, the relative location of the object's individual components being of a key importance. For instance, a cup and a pail are both comprised of identical geons—a curved handle and a cylinder, yet they are easily distinguishable from each other. The number of possibilities of joining individual shapes observed in parallel can be infinite (Biederman 1987).

Biedermann's theory was repeatedly supported in other studies presenting pigeons e.g. with drawings whose lines were incomplete (Van Hamme et al. 1992) or were made up of a small number of geons (Wasserman et al. 1993). A different procedure for testing the RBC was opted for by Kirkpatrick-Steger et al. (1996). Pigeons were trained to recognize four basic geometric shapes—cylinder, cone, handle and wedge. All the shapes were shown to them together with a cube, in different spatial organization. During the training, the birds learnt to differentiate these pairs of shapes from negative stimuli. The speed of their response depended on the content of the negative stimulus—if it contained the same shape as the positive stimulus, the birds responded more slowly. Consequently, Kirkpatrick-Steger et al. (1996) concluded that even though all birds were able to learn not to recognize stimuli based on the location of the features, they still paid more attention to their mere presence, taking the location into consideration only in the second place.

Studies testing the RBC method mentioned above used very specific stimuli. These were simple drawings composed of more or less general geometric shapes (geons in the sense defined by Wasserman et al. 1993). Therefore, the tested pigeons did not have any other information than their shape and relative location for their discrimination. Real objects (including their photos and drawings) are much more complex; more or less unique elements (features) as well as their relative location can be used for their discrimination, and either of them can be chosen.

Watanabe (2001a) worked with such stimuli in order to resolve the contradiction between studies supporting the PFT and RBC. He carried out a total of four experiments differing in the realness of the used stimuli and their relevance for the tested pigeons. In the training, the first group of pigeons was shown colourful drawings of ten heroes of a Japanese comic book, while the second group of pigeons learnt to discriminate black and white pigeon drawings from drawings of other birds. The third and fourth groups were taught to recognize people or birds in photos. The success rate of individual groups of pigeons in discriminating mixed objects significantly differed among groups working with photographs and images. Whilst with the images the success rate almost did not drop, it was considerably lower compared to the training in the case of the photos. Ergo, the experiments with images more likely supported the PFT theory, whereas the experiments with photos supported the RBC theory.

In addition to the realistic manner of the presented stimuli, a significant effect of their overall complexity on the success of recognition was ascertained. In the experimental phase, Matsukawa et al. (2004) presented pigeons with drawings of two comic characters geometrically split (Fig. 4.2). With an increasing extent of fragmentation, the pigeons showed a lower recognition ability; probably for the reason that partial stimuli gradually disappear. Presenting photographs of people, objects and animals by Ullman et al. (2002) demonstrated that if the stimulus is too detailed (a small section of an object) or contrariwise too complex (a fragmented object, an overall view of a building, a group of people), the recognition efficiency is conclusively lower than if medium-complex stimuli are presented (a face, a section of a car, etc.).

From the point of view of behavioural ecology, probably the most interesting experiments were carried out by Cook et al. (2013), who trained pigeons to differentiate black and white drawings of birds and mammals, i.e. objects bearing similar potential cues as ecologically relevant stimuli. In the experiment, the images were split into three parts (head, body, legs) and a number of modifications were made from them testing what significance the individual parts of the body (chimeras) and the integrity and arrangement of the drawings (more or less separated and mixed parts, Fig. 4.3) had for distinguishing. The head was crucial for discriminating birds and mammals, while the importance of the body and legs was significantly lower. Changing the number of feet (2 vs. 4) had no significant effect, either. The success rate of recognition was not considerably reduced by mixing the individual parts—not even if they were separated by a gap as well. Taking into account the results of Matsukawa et al. (2004), this is not entirely surprising. The stimuli of Cook et al. (2013) were only split into three parts, thus keeping all potential local features in an unchanged form. However, the stimuli disintegration having a minimal effect constitutes a new finding.

Nonetheless, each separate part of the body in this study represents a complex object even in the used black and white drawings. It can be broken down into a number of simple geometric shapes and besides that, it is endowed with a unique internal structure. Therefore, these again are not geons as defined by Wasserman et al. (1993). That may be another reason why Cook et al. (2013) speak about local (body parts) and global (relative location of the body parts) features. In our opinion,

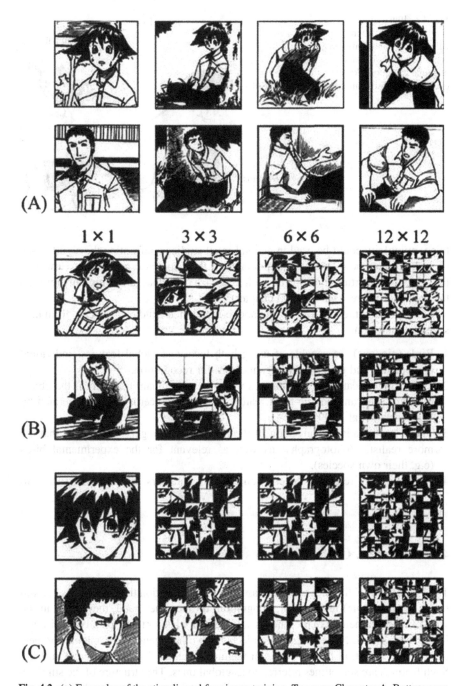

Fig. 4.2 (a) Examples of the stimuli used for pigeon training. Top row: Character A. Bottom row: Character B. (b) Selections of novel stimuli (1 × 1) and thein fragmented versions of free different degrese (3 × 3, 6 × 6, and 12 × 12). (c) Novel stimuli depicting close-up faces and the fragmented versions (Matsukawa et al. 2004 with permission)

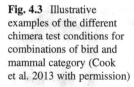

Fig. 4.3 Illustrative examples of the different chimera test conditions for combinations of bird and mammal category (Cook et al. 2013 with permission)

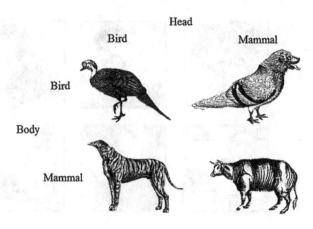

this terminology could be used in general in all studies confronting the significance of partial components with the importance of their spatial organization. However, the 'global' properties of an object, such as its size or surface, are commonly referred to as global features (see the next chapter).

Summarizing the results of the above-mentioned studies, we can formulate several conclusions.

1. Birds (pigeons) are capable of using both local and global features (individual components and their spatial organisation) for recognition.
2. The relative location of the features will play a dominant role only if the object does not offer any unique local features, i.e. if it concerns geons as defined by Wasserman et al. (1993).
3. The role of spatial organisation of the features is stronger with objects that are more realistic (photographs) as well as relevant for the experimental birds (e.g. their own species).
4. Regarding the relevant objects, some local characters are preferred (head) to others (torso).

4.2   Local vs. Global Features

The PFT and RBC theories are both based on the assumption that animals use partial (more likely simple) features for recognition; on the other hand, they differ in the view of whether their mere presence is sufficient or whether their spatial organisation is relevant as well. Nonetheless, all objects have other 'general' or 'global' properties that are not at all bound to partial features, such as size or a different surface. Such features are sometimes referred to as global ones. The structure of the stimulus determined by the relative location of individual local features can be, however, included among the global features as well (Cerella 1986).

The significance of the size and general shape of the stimulus was tested in several studies. Troje et al. (1999) showed that pigeons prefer internal features of the face (surface structure, hue, brightness) to general shape and size when categorizing the human face. Yet if the photographs were modified to have the same internal features, the pigeons were able to make use of the size and shape of the face for discrimination. Aust and Huber (2001, 2002) came to similar conclusions in experiments with photographs of human figures or parts of them.

On the contrary, the experiment of Lazareva et al. (2006) demonstrated substantial importance of the general shape for recognising a human figure, while the loss of information about its structure by blurring the photograph did not affect the discrimination. As opposed to the stimuli used by Aust and Huber, the stimuli in this study contained no background that could disturb the good visibility of a body silhouette; subsequently it was easier to categorize on the basis of the object's global features.

Further studies focused on objects with a homogeneous internal texture repeatedly demonstrated that pigeons are capable of using it for the recognition of artificial stimuli (e.g. Katz and Cook 2000; Young et al. 2001; Kelly and Cook 2003). However, there are other works showing that the importance of the surface texture failed to outweigh the importance of the shape in recognizing, namely in the case of presenting man-made objects (e.g. automobiles—Nicholls et al. 2011). This can be caused by a considerably higher variability in the shapes of cars and houses than, for example, in the shapes of faces.

4.3 Category Discrimination and Concepts Formation

So far we have used the term recognition in the text. At the beginning of this chapter it must be mentioned that the terminology in the field of behavioural ecology and ethology is not identical with the terminology of psychological and cognitive sciences. Cognitive sciences more related to human psychology perceive the word recognition as a process describing the 'recalling' of previously acquired knowledge, e.g. of the identity of the predator. On the other hand, behavioural sciences use the term recognition for identifying a noticed animal as a predator, i.e. differentiating a predator from a harmless animal, or distinguishing predators from one another (Shettleworth 2010), without looking into the recognition mechanism. In the terminology of psychology, the process of differentiating a predator from a harmless animal, or alternatively various predators from one another, can be called 'discrimination' (Herrnstein et al. 1976). However, recognition also requires having a particular sets of objects created in some way (e.g. predators/non-predators, corvids/predators). Psychology calls such groups 'categories' and the process of creating them 'categorization' (Pearce 2008).

Actually, the idea that birds (and other animals) actively form categories is quite a new one, being to a large extent contrary to the concept of classical ethologists. They assumed that responses to surrounding objects, e.g. by predators, are mostly instinctive and similar to simple reflexes (Tinbergen 1951). Just as a lot of light constitutes a stimulus for the pupil, spotting a curved bird beak triggers escape. It was the American psychological school that began experimentally studying the process of creating categories in animals.

The experiment by Herrnstein and Loveland (1964) represented a pioneering work testing the ability of birds to categorize. In the experiment, they used colour photographs showing different situations, environments and objects. These photographs were divided into two categories (human/non-human) depending on whether a human figure was present in them or not (one or more). The pigeons successfully learnt to differentiate between these two categories as well as demonstrating the ability to generalize the categorization to new photographs of both categories. Pigeons' ability to form a concept of other categories was tested in further experiments (tree, water surface, fish, or one particular person—Herrnstein et al. 1976; Herrnstein and de Villiers 1980; Vaughan and Herrnstein 1987). It always concerned the categorization of very variable and complex photographs whose affiliation with one of two categories was only determined by the presence of a particular object.

Besides natural objects, pigeons can also categorize artificial, man-made objects (chairs, cars—Bhatt et al. 1988 or Lazareva et al. 2004) and even at first glance very comprehensively specified categories like Impressionism and Cubism (Watanabe et al. 1995; Watanabe 2001b). These experiments are very impressive; nonetheless, they fail to show how pigeons proceed when creating categories.

4.4 Features × Exemplars × Prototypes

There are several theories describing how animals can form categories. An animal can have some congenital categories; others are developed individually from experience with its representatives throughout their life (Pearce 1994).

The feature theory assumes the existence of a set of simple (unambiguous) structures that clearly define a stimulus belonging to a particular category (Bruner et al. 1956; Huber 2001). The set of features is specific for each category. A new stimulus is categorized by comparing the features it contains with the features defining the category. Such a procedure does not contradict the conceptions of classical ethology about the recognition of ecologically significant objects and the triggering of corresponding behaviour (Tinbergen 1951). From the point of view of cognitive sciences, it is actually not a matter of categorization, as there is no need to use generalization. The fact that individual objects differ in some features is not important; it is sufficient if they have the features defining the respective category.

The exemplar theory assumes the maintaining of all the representatives of a category the animal has already encountered in memory (Astley and Wasserman 1992; Huber and Lenz 1993; Huber 2001). Categorization takes place by comparing the stimulus with the memorized exemplars. The more the stimulus is similar to the memorized exemplars, the better the categorization is. The need to remember an enormous amount of stimuli that the animal has encountered constitutes an evident demerit of this theory. However, some studies (Herrnstein et al. 1976; Vaughan and Greene 1984) show that pigeons are capable of remembering thousands of images.

The prototype theory derived from human psychology (Posner and Keele 1968) assumes the existence of a typical representative of a category—an 'average exemplar' that does not necessarily exist itself in the real world (Neumann 1977; Huber 2001). Subsequently, the categorization of the stimulus then proceeds by comparing it with the prototype of the category and, like in the case of the exemplar, the more similar the categorized stimulus is to the prototype the more successful the categorization is. This theory was supported by the fact that exemplars more resembling the supposed prototype are categorized more simply by humans than more unalike exemplars (Shanks 1994). The prototype theory assumes the abstraction ability so that the non-existent average exemplar of the particular category can be formed, as well as the generalization ability so that the existing exemplars can be compared with the average exemplar. However, there are experimental studies supporting this theory (Von Fersen and Lea 1990; Aydin and Pearce 1994).

Human psychology uses the concept of conceptualization for the formation of a mental representation of a typical representative of a particular category, calling its result the concept of a category. It can be discussed to what extent the concept is identical to the prototype; nonetheless, in most of the above-mentioned studies, this terminology is also adopted into pigeon psychology (see review Jitsumori 2004). However, these days a trend has been emerging where the concept formation is placed above the mere ability to categorize similar stimuli differing in local or global features. As the concept formation is subsequently considered a situation where a particular category is formed really universally and is recognized even when e.g. more stimuli are presented or in different spatial organisation (Zorina and Obozova 2011; Obozova et al. 2012). Studies concerning species of birds with a higher brain complexity, such as corvids (Smirnova et al. 2003) and parrots (Suková et al. 2013), usually assume that the birds have demonstrated the concept formation in a meaning similar to that of higher primates and humans. Nonetheless, for example jungle crows (*Corvus macrorhynhcus*) taught to differentiate a triangle and a square considered as a triangle all triangles differing in colour, length and rations of the sides—even a mere triangle outline, yet a change in the triangle size significantly diminished the categorization ability (Bogale and Sugita 2014). In this case we should not refer to it directly as the concept formation.

4.5 Functional Categorization

All of the studies described hereinbefore studied the ability of pigeons to form so-called perceptual categories (Shettleworth 2010), which are defined by directly perceivable properties of the stimuli, therefore in the case of all the above-mentioned pigeon experiments by their appearance. Applied to predator recognition, a predator is defined by visual features characteristic only just of this group (a curved beak, talons, etc.). Besides this simplest level of categorization, human psychology in particular recognizes the so-called functional categorization, where the functional properties of the stimulus define the categories, rather than its appearance. Again when applied to predators, a raptor is recognized as a predator, i.e. a life-threatening animal. Functional categorization is more demanding at first sight than perceptual categorization—it requires a higher degree of abstraction and generalization, as it forms categories based on visually (or more generally sensorially) non-manifested properties. Wasserman et al. (1993) and later Lazareva et al. (2004) assume that their experiments demonstrate the ability of animals to form functional categories.

In the experiment by Wasserman et al. (1993), photographs of four groups of objects (chairs, cars, human figures, flowers) were presented to pigeons. During the training, they were taught to divide these objects into two groups, therefore responding in the same way to two different categories of objects. They were then reoriented to respond to one of these categories differently. In the experiment itself, both categories were presented to them, with the pigeons responding to the original category identically as to the re-educated one with a higher-than-accidental probability. Better results were achieved if the two categories consisted of objects of a same origin, i.e. human figures and flowers, or chairs and cars, rather than their combinations, i.e. for instance figures and chairs, or flowers and cars. It is this result that the authors interpret as proof that the pigeons formed functional categories of artificial and natural objects. However, there could be another explanation. The pigeons might have only been guided by the similarity of the tested objects. Both flowers and human figures are characterized e.g. by the predominance of rounded edges, while straight edges dominate in chairs and cars. A study by Lazareva et al. (2004) convincingly demonstrates that pigeons are capable of creating hierarchical categories (from the same objects as used by Wasserman et al. 1993), which undoubtedly constitutes a prerequisite for the ability to form functional categories. Admittedly, this again does not mean that the pigeons used the functional similarity of the used objects when performing the task.

4.6 Process of Discrimination and Categorization in Behavioural Experiments with Untrained Birds

4.6.1 Cues Used for Discrimination and Categorization

Perhaps the most important piece of knowledge gained in experiments with untrained birds is that the cues used in predator recognition are not unequivocally determined. Thus, the concept of the classical European ethological school does not apply (or at least not in full), i.e. that predator recognition works on the reflex principle (Tinbergen 1951) when a simple stimulus (trigger)—in the case of predators the presence of a key feature or features, automatically causes an appropriate response—anti-predatory behaviour in the case of predators.

Nonetheless, experiments with untrained birds abundantly demonstrate that noticeable partial features unique to a particular group or even individual species are used for recognizing predators. Removing such features or replacing them with features of harmless animals often results in the predator not being recognized (e.g. Curio 1975). On the other hand, the very presence of such features on a harmless animal is usually insufficient for recognizing it as a predator (Beránková et al. 2014). Besides the presence of the key features, the context in which they occur is also important for recognition. Most probably, its role may increase if it is functionally important, as shown by the experiments with recognising a sparrowhawk and a cuckoo (Welbergen and Davies 2008). However, what the necessary context really involves is quite unclear. Recognition can be prevented by a wrong location of a key feature on the otherwise unchanged stimulus (e.g. a black eye stripe located on the chest instead of the head), an incorrect look of the feature (e.g. a colour-wise distinctly different eye stripe), or even an incorrect overall appearance (e.g. a significantly different colouring—Curio 1975).

Regrettably, the studies on pigeons did not use any changes testing the necessity or sufficiency of partial features for recognition. A study by Cook et al. (2013) approximates this type of tasks most, using the exchange of individual body parts body in animal drawings apart from other things. The result of this experiment is extremely interesting for behavioural ecology. To distinguish between birds and mammals, the head was crucial, which is consistent with the results of a wide range of studies testing untrained birds regarding the importance of the eyes, beak and the partial components of the colouring (Sects. 3.2.3 and 3.2.4). This would imply that birds are most likely taught by a contact with their parents or possibly other conspecific individuals—or it is congenital for them—that the features for recognizing other animals can mainly be provided by the head. This again does not fully correspond to the concept of classical ethology about the isolated function of key features.

On the other hand, the experiments with untrained birds paid little attention to the importance of the overall organisation of partial features, i.e. to the RBC and PFT hypotheses, relatively intensively tested on pigeons. It was Nováková et al. (2017) who carried out an experiment fully corresponding to some of these studies on pigeons regarding the organisation. They presented birds visiting a feeder with a model of a Eurasian sparrowhawk divided into three parts, which were mixed in various ways. Since this experiment was carried out in a two-feeder arrangement (Chap. 1), the authors could show that the birds prefer to fly to the pigeon rather than to all sparrowhawk models (unmodified and modified), and also that if they have a choice between an unmodified and a modified sparrowhawk they prefer not to fly anywhere, perceiving them both as equivalent threats. This fully corresponds to the conclusions of Cook et al. (2013), who worked with drawings of birds and mammals, as well as to a classic study by Cerella (1986) with the pictures of Charlie Brown. Undoubtedly, this result is surprising, since from a human point of view the model with mixed parts considerably differs from the unmodified sparrowhawk. However, the results of Nováková et al. (2017) are different from the conclusions of Curio's experiments (1975), in which flycatchers took the relative location of the colouring components into consideration. Nonetheless, he only changed the position of one feature (eye stripe), while in the case of Nováková et al. (2017) the units bearing potential key features (head, body, legs) remained unchanged.

Classic experiments with silhouettes of predators (e.g. Krätzig 1940) are also related to the issue of the RBC theory vs. PFT theory. These stimuli do not provide any unique partial features. The tested birds recognized them on the basis of the relative location of individual body parts that correspond to the definition of geons by Wasserman et al. (1993). If the silhouette moves with its tail forward, it actually means that the long tail and short head end up in the wrong place and the silhouette is not recognized as a predator. As far as experiments with pigeons are concerned, a study by Wasserman et al. (1993) where pigeons failed to recognize simple drawings of artificial objects with mixed parts resembles these experiments most. Even untrained birds thus have the ability to use the organisation of the object's partial components for recognition. Nonetheless, as well as trained birds they mainly use it when the object (a flying bird in their case) does not provide any unique partial features.

If we summarize the current knowledge of predator recognition by untrained birds, we can see that they are more in concordance with the PFT theory, nevertheless not in absolute correspondence. It seems to be well supported with evidence that to recognize predators birds use, with the exception of silhouettes, distinct partial features, which can be termed as key features in traditional terminology. On the contrary, the importance of their relative location was not confirmed, but in this case the existing evidence is very fragmentary.

However, the key features on their own are not sufficient for recognition; they must occur in a certain context, but then the 'nature' of this context is unclear.

4.6.2 Category Formation

The existing studies focused on predator recognition by untrained birds collected in Chap. 3 undoubtedly show that birds are able not only to distinguish predators from harmless birds but also groups or individual types of predators differing in the danger they pose to the tested species. At first sight, it can thus be inferred that they formed a hierarchical system of categories: e.g. Eurasian sparrowhawk, daytime raptor, predator. According to some authors, this system requires conceptualization ability (Jitsumori 2004).

Nonetheless, in reality, these results do not give us evidence how birds form these categories. A zero hypothesis is undoubtedly constituted by the fact that they only form categories based on the presence of key features: e.g. a yellow eye and horizontal barring of the underparts = sparrowhawk; a curved beak, eyes with a supraorbital ridge and long curved claws = a daytime raptor; sharp weapons (beak, claws) = predator. Adequate responses to individual species or groups of predators thus do not necessarily require any generalization, not to mention conceptualization.

As summarized in the previous chapter, the results of a number of studies show that predator recognition simply does not work this way. Though key features are usually of a key importance for recognition, they must occur in a certain context. This should mean that the respective—more or less general—categories are defined more comprehensively than by their mere presence.

The ability of birds to reliably distinguish predators on a generic level, including species very similar in some features, could be grounded on their very good memory, whereas categories would be defined on an exemplar principle. During an encounter, the bird would compare the relevant predator with those met during its lifetime. Experimental studies on pigeons showed that they manage to learn and remember hundreds of individual photos in the long term (Herrnstein et al. 1976; Vaughan and Greene 1984). It can be assumed that this ability originated due to the need to reliably recognize relevant objects (including predators). However, the exemplar theory can hardly be applied to experiments in which unknown predators were presented to birds, including chimeras of dangerous and harmless species (Beránková et al. 2014). Experiments by Veselý et al. (2016) are perhaps most interesting. A sparrowhawk coloured like a European Robin or a sparrowhawk without the horizontal barring on its stomach triggered the same fear in tits on a feeder as an unmodified sparrowhawk. This result could be attributed to the fact that these modifications retained other features characterizing the sparrowhawk (yellow eyes in combination with a raptor beak, claws, size and body shape). However, the last model with violet-white chequer colouration was also similarly modified, yet it did not cause any fear. If the above-mentioned features were sufficient for being included in the sparrowhawk category, this model should have been included in it as well. It seems that the categorization process in this case was more complex. In the first stage, a classification into a broad category took place (probably a bird). An artificially coloured model failed to 'pass' this process, while other models with realistic colouring known to the tested birds (robin, sparrowhawk) passed. It is difficult to imagine that such a category could be formed on the basis of the presence of simple features as well as through comparing with all the birds that the tested tits have encountered, as this would require extraordinary memory indeed. What appears to be more likely is that it was formed on the basis of a prototype bearing general bird features and the geometric chequer colouration did not fit in it. It was only in the second stage that the models were classified in the sparrowhawk category; probably owing to the presence of simple features, which more or less unequivocally characterized it.

The existence of generally defined categories is also suggested by the experiment of Němec and Fuchs (2014), in which shrikes attacked an unknown Spotted Nutcracker as often as a known jay, despite the nutcracker being harmless. The authors suggest that the nutcracker was classified in the corvid bird category. The corvids are predominantly predators of nests; therefore, it is necessary to attack them in the proximity of the nest. A somewhat different indirect evidence of the ability of birds to conceptualize is provided by the study of Tvardíková and Fuchs (2012), in which they proved that birds are capable of amodal completion, i.e. they 'know' that a partially hidden object (in their case, a sparrowhawk model) is incomplete. This, however, requires a previous formation of an 'abstract idea' of the sparrowhawk, for which we could possibly use the term 'concept'.

4.7 Future Research Vision

Predator recognition is an essential ability of all animals; therefore, it is also useful in studying cognitive processes in birds. Existing studies, starting with classical etho-logical experiments, have provided convincing evidence that birds use characteristic partial features when recognizing.

However, these features only work in a certain context. So far this context has been just fragmentary. In our opinion, there is an interesting field for experimental research here. Some of the studies performed on pigeons could probably provide some inspiration. As we saw in the previous chapters, they dealt with the properties of the objects to be recognised that behavioural ecology had neglected until then, as they do not apply to environmentally relevant stimuli in the nature. A potential prey, for example, cannot encounter a predator that has 'mixed' features. However, this does not mean that there is no point in questioning whether or not the organization of partial features affects predator recognition.

Questions that come forward for research carried out on untrained birds can be divided into several areas: (1) What is the role of 'global' features of the object to be recognised? This includes its properties such as size, shape, location or movement, but also the relative location of the partial features (RBC vs. PFT), including the extent of their integration. (2) What is the relationship of individual key features? We can ask whether there is a hierarchy between them; whether the absence of individual key features reduces the ability to recognize the object, or whether the consequences of the absence of a feature and its substitution by an 'incorrect' feature differ. (3) What is the relationship between the key and other partial features of the object? In the particular case of bird predators, it is the colouring, i.e. the hue, texture and arrangement of partial areas.

By answering the above-mentioned questions, we should at least clarify our knowledge of what categories birds form in predator recognition, and how these categories are defined. A much more challenging task is to study the way birds assign new objects to the categories formed. The existing experiments show that they are not only guided by the presence of key features. However, this begs a question whether they rather compare the new object with previously known exemplars of a particular category, or whether they formed a generalized prototype for it.

Experiments based on discrimination learning could bring new findings, which would allow controlling the category forming process and subsequently modifying the tested exemplars as well as comparing how different types of modifications, such as removing features, replacing them with new features, combining features from different categories, or globally modifying the object will change the ability of birds to assign the modified exemplars to categories they have been taught. This could be different for the categories formed on the basis of already known exemplars and categories represented by a generalized prototype.

Just as behavioural ecology did not address some of the properties of the objects to be recognised, behavioural psychologists mostly avoided working with environ-mentally relevant stimuli. In the narrow sense of the word, such natural objects as trees or human figures cannot be considered. This is quite understandable as in experiments based on discrimination learning it is not desirable for results to be affected by congenital knowledge, which must be assumed for environmentally relevant stimuli.

However, working with environmentally relevant knowledge may probably bring interesting pieces of knowledge. Cook et al. (2013) tested reactions to modified animal drawings. Despite using pretty schematic line-drawings, the results suggest that the pigeons recognized animals and that they view the animals in a specific way. Most experiments based on discrimination learning work with domestic pigeons. They have proved to be a very good co-operating species, but the question is to what extent the obtained results can be generalized to other birds. There are also other species whose ability to adapt to laboratory conditions is quite good, e.g. tits (Zorina and Obozova 2011). In addition, it would be possible to simulate a predation context. Birds would be trained to fear certain stimuli, and subsequently these stimuli would be modified in different ways.

Behavioural manifestations of the tested birds do not directly indicate whether they recognized the presented predator, but rather whether they evaluated it as an imminent danger (see Chap. 1), which constitutes a relatively significant complication of experiments with untrained birds. Therefore, it is advisable to test the same stimuli in both nesting and feeder experiments. In the first mentioned experiments, the birds only respond by intense antipredator behaviour if they are confident in their recognition, in the latter experiments they are more guided by the precautionary principle. Nonetheless, the use of physiological markers in cage or aviary experiments appears to head in the most promising direction.

References

Abrahams MV, Dill LM (1989) A determination of the energetic equivalence of the risk of predation. Ecology 70:999–1007. https://doi.org/10.2307/1941368

Ackerman JT, Takekawa JY, Kruse KL, Orthmeyer DL, Yee JL, Ely CR, Ward DH, Bollinger KS, Mulcahy DM (2004) Using radiotelemetry to monitor cardiac response of free-living tule greater white-fronted geese (*Anser albifrons* elgasi) to human disturbance. Wilson Bull 116:146–151. https://doi.org/10.1676/03-110

Adams JL, Camelio KW, Orique MJ, Blumstein DT (2006) Does information of predators influence general wariness? Behav Ecol Sociobiol 60:742–747. https://doi.org/10.1007/s00265-006-0218-9

Adriaensen F, Dhont AA, Van Dongen S, Lens L, Matthysen E (1998) Stabilizing selection on blue tit fledgling mass in the presence of sparrowhawks. Proc R Soc Lond B 265:1011–1016. https://doi.org/10.1098/rspb.1998.0392

Alatalo RV, Helle P (1990) Alarm calling by individual willow tits, *Parus montanus*. Anim Behav 40:437–442. https://doi.org/10.1016/S0003-3472(05)80523-8

Albano N, Santiago-Quesada F, Masero JA, Sanchez-Guzman JM, Mostl E (2015) Immunoreactive cortisone in droppings reflect stress levels, diet and growth rate of gull-billed tern chicks. Gen Comp Endocr 213:74–80. https://doi.org/10.1016/j.ygcen.2015.02.019

Altmann SA (1956) Avian mobbing behavior and predator recognition. Condor 58:241–253. https://doi.org/10.2307/1364703

Amat JA, Masero JA (2004) Predation risk on incubating adults constrains the choice of thermally favourable nest sites in a plover. Anim Behav 67:293–300. https://doi.org/10.1016/j.anbehav.2003.06.014

Amat JA, Carrascal LM, Moreno J (1996) Nest defence by Chinstrap Penguins *Pygoscelis antarctica* in relation to offspring number and age. J Avian Biol 27:177–179. https://doi.org/10.1016/j.anbehav.2003.06.014

Amo L, Galván I, Tomás G, Sanz JJ (2008) Predator odour recognition and avoidance in a songbird. Funct Ecol 22:289–293. https://doi.org/10.1111/j.1365-2435.2007.01361.x

Amo L, López-Rull I, Pagán I, García CM (2015) Evidence that the house finch (*Carpodacus mexicanus*) uses scent to avoid omnivore mammals. Rev Chil Hist Nat 88. https://doi.org/10.1186/s40693-015-0036-4

Andersson M, Wiklund CG, Rundgren H (1980) Parental defence of offspring: model and an example. Anim Behav 28:536–542. https://doi.org/10.1016/S0003-3472(80)80062-5

Apel KM (1985) Antipredator behavior in the Blackcapped Chickadee (*Parus atricapillus*). Dissertation, University of Wisconsin

Arnold KE (2000) Group mobbing behaviour and nest defence in a cooperatively breeding Australian bird. Ethology 106:385–393. https://doi.org/10.1046/j.1439-0310.2000.00545.x

© The Author(s) 2019

R. Fuchs et al., *Predator Recognition in Birds*, SpringerBriefs in Animal Sciences, https://doi.org/10.1007/978-3-030-12404-5

Arnold JM, Ordonez R, Copeland DA, Nathan R, Scornavacchi JM, Tyerman DJ, Oswald SA (2011) Simple and inexpensive devices to measure heart rates of incubating birds. J Field Ornithol 82:288–296. https://doi.org/10.1111/j.1557-9263.2011.00332.x

Arroyo B, Mougeot F, Bretagnolle V (2001) Colonial breeding and nest defence in Montagu's harrier (*Circus pygargus*). Behav Ecol Sociobiol 50:109–115. https://doi.org/10.1007/s002650100342

Ash J (1970) Observations on a decreasing population of red-backed shrikes. Br Birds 63:185–205

Astheimer LB, Buttemer WA, Wingfield JC (1994) Gender and seasonal differences in the adreno-cortical response to ACTH challenge in an Arctic passerine, *Zonotrichia leucophrys gambelii*. Gen Comp Endocrinol 94:33–43. https://doi.org/10.1006/gcen.1994.1057

Astley SL, Wasserman EA (1992) Categorical discrimination and generalization in pigeons: all negative stimuli are not created equal. J Exp Psych Anim Behav Process 18:193–207. https://doi.org/10.1037/0097-7403.18.2.193

Aust U, Huber L (2001) The role of item- and category-specific information in the discrimination of people versus nonpeople images by pigeons. Anim Learn Behav 29:107–119. https://doi.org/10.3758/BF03192820

Aust U, Huber L (2002) Target-defining features in a "people-present/people-absent" discrimination task by pigeons. Anim Learn Behav 30:165–176. https://doi.org/10.3758/BF03192918

Avilés JM, Parejo D (2006) Nest defense by Iberian azure-winged magpies (*Cyanopica cyanus*): do they recognize the threat of brood parasitism? Ethol Ecol Evol 18:321–333. https://doi.org/10.1080/08927014.2006.9522699

Aydin A, Pearce JM (1994) Prototype effects in categorization by pigeons. J Exp Psych Anim Behav Process 20:264–277. https://doi.org/10.1037/0097-7403.20.3.264

Azevedo CS, Young RJ (2006a) Behavioural responses of captive-born greater rheas *Rhea americana* Linnaeus (Rheiformes, Rheidae) submitted to antipredator training? Rev Bras Zool 23:186–193. https://doi.org/10.1590/S0101-81752006000100010

Azevedo CS, Young RJ (2006b) Do captive-born greater rheas *Rhea americana* Linnaeus (Rheiformes, Rheidae) remember antipredator training? Rev Bras Zool 23:194–201. https://doi.org/10.1590/S0101-81752006000100011

Azevedo CS, Rodrigues LSF, Fontenelle LCR (2017) Important tools for Amazon Parrot reintroduction programs. Rev Brasil Ornitol 25:1–11

Baker MC, Becker AM (2002) Mobbing calls of black-capped chickadees: effects of urgency on call production. Wilson Bull 114:510–516. https://doi.org/10.1676/0043-5643(2002)114[0510:MCOBCC]2.0.CO;2

Balda RP, Bateman GC (1973) Unusual mobbing behavior by incubating Pinon Jays. Condor 75:251–252. https://doi.org/10.2307/1365882

Barash DP (1975) Evolutionary aspects of parental behavior: distraction behavior of the Alpine Accentor. Wilson Bull 87:367–373

Barash DP (1976) Mobbing behavior by crows: the effect of the "crow-in-distress" model. Condor 70:120. https://doi.org/10.2307/1366937

Bártol I, Karcza Z, Moskát C, Røskaft E, Kisbenedek T (2002) Responses of great reed warblers *Acrocephalus arundinaceus* to experimental brood parasitism: the effects of a cuckoo *Cuculus canorus* dummy and egg mimicry. J Avian Biol 33:420–425. https://doi.org/10.1034/j.1600-048X.2002.02945.x

Bautista LM, Lane SJ (2000) Coal tits increase evening body mass in response to tawny owl calls. Acta Ethol 2:105–110. https://doi.org/10.1007/s102119900014

Bažant M (2009) How birds judge the risk of predation during winter nourishment experiments – meaning of the movement of the model. BSc thesis in University of South Bohemia, Czech

Bazin RC, Sealy SG (1993) Experiments on the responses of a rejector species to threats of predation and cowbird parasitism. Ethology 94:326–338. https://doi.org/10.1111/j.1439-0310.1993.tb00449.x

Beauchamp G (2001) Should vigilance always decrease with group size? Behav Ecol Sociobiol 51:47–52. https://doi.org/10.1007/s002650100413

Beránková J, Veselý P, Sýkorová J, Fuchs R (2014) The role of key features in predator recognition by untrained birds. Anim Cogn 17:963–971. https://doi.org/10.1007/s10071-014-0728-1

Beránková J, Veselý P, Fuchs R (2015) The role of body size in predator recognition by untrained birds. Behav Process 120:128–134. https://doi.org/10.1016/j.beproc.2015.09.015

Betts MG, Halley AS, Doran PJ (2005) Avian mobbing response is restricted by territory boundaries: experimental evidence from two species of forest warblers. Ethology 111:821–835. https://doi.org/10.1111/j.1439-0310.2005.01109.x

Bhatt RS, Wasserman EA, Reynolds WFJ, Knauss KS (1988) Conceptual behavior in pigeons: categorization of both familiar and novel examples from four classes of natural and artificial stimuli. J Exp Psych Anim Behav Process 14:219–234. https://doi.org/10.1037/0097-7403.14.3.219

Biederman I (1987) Recognition-by components: a theory of human image understanding. Psychol Rev 94:115–147. https://doi.org/10.1037/0033-295X.94.2.115

Binazzi R, Zaccaroni M, Nespoli A, Massolo A, Dessi-Fulgheri F (2011) Anti-predator behaviour of the red-legged partridge Alectoris rufa (Galliformes: Phasianidae) to simulated terrestrial and aerial predators. Ital J Zool 78:106–112. https://doi.org/10.1080/11250003.2010.509136

Blancher PJ, Robertson RJ (1982) Kingbird aggression: does it deter predation? Anim Behav 30:929–930. https://doi.org/10.1016/S0003-3472(82)80167-X

Bogale BA, Sugita S (2014) Shape discrimination and concept formation in the jungle crow (Corvus macrorhynchos). Anim Cogn 17:105–111. https://doi.org/10.1007/s10071-013-0642-y

Bogrand AL, Neudorf DLH, Matich P (2017) Predator recognition and nest defence by Carolina Wrens Thryothorus ludovicianus in urban and rural environments: does experience matter? Bird Study 64:211–221. https://doi.org/10.1080/00063657.2017.1316235

Borkin LY, Litvinchuk SN, Rozanov YM, Skorinov DV (2004) On cryptic species (from the example of amphibians). Zool Zh 83:936–960

Bosque C, Molina C (2002) Communal breeding and nest defense behavior of the Cayenne Jay (Cyanocorax cayanus). J Field Ornithol 73:360–362. https://doi.org/10.1648/0273-8570-73.4.360

Breuner CW, Delehanty B, Boonstra R (2013) Evaluating stress in natural populations of vertebrates: total CORT is not good enough. Funct Ecol 27:24–36. https://doi.org/10.1111/1365-2435.12016

Briskie JV, Sealy SG (1989) Changes in nest defense against a brood parasite over the breeding cycle. Ethology 82:61–67. https://doi.org/10.1111/j.1439-0310.1989.tb00487.x

Brown CR, Hoogland JL (1986) Risk in mobbing for solitary and colonial swallows. Anim Behav 34:1319–1323. https://doi.org/10.1016/S0003-3472(86)80203-2

Bruner JS, Goodnow J, Austin F (1956) A study of thinking. Wiley, New York, NY

Brunton DH (1986) Fatal antipredator behavior of killdeer. Wilson Bull 98:605–607

Brunton DH (1990) The effects of nesting stage, sex, and type of predator on parental defense by Killdeer (Charadrius vociferous): testing models of avian parental defense. Behav Ecol Sociobiol 26:181–190. https://doi.org/10.1007/BF00172085

Brunton DH (1997) Impacts of predators: center nests are less successful than edge nests in a large nesing colony of lest terns. Condor 99:372–380. https://doi.org/10.2307/1369943

Brunton D (1999) "Optimal" colony size for least terns: an intercolony study of opposing selective pressures by predators. Condor 101:607–615. https://doi.org/10.2307/1370190

Buitron D (1983) Variability in the responses of Black-billed Magpies to natural predators. Behaviour 78:209–236. https://doi.org/10.1163/156853983X00435

Bump SR (1986) Yellow-headed blackbird nest defense: aggressive responses to marsh wrens. Condor 88:328–335. https://doi.org/10.2307/1368880

Bureš S, Pavel V (2003) Do birds behave in order to avoid disclosing their nest site? Bird Study 50:73–77. https://doi.org/10.1080/00063650309461293

Burger J, Gochfeld M (1992) Experimental evidence for aggressive antipredator behavior in Black Skimmers (Rynchops niger). Aggressive Behav 18:241–248. https://doi.org/10.1002/1098-2337(1992)18:3<241::AID-AB2480180307>3.0.CO;2-3

Burger J, Gochfeld M, Saliva JE, Gochfeld D, Gochfeld D, Morales H (1989) Antipredator behaviour in nesting zenaida doves (Zenaida aurita): parental investment or offspring vulnerability? Behaviour 65:129–143. https://doi.org/10.1163/156853989X00628

Burger J, Shealer DA, Gochfeld M (1993) Defensive aggression in terns: discrimination and response to individual researchers. Aggressive Behav 19:303–311. https://doi.org/10.1002/1098-2337(1993)19:4<303::AID-AB2480190406>3.0.CO;2-P

Burhans DE (2000) Avoiding the nest: responses of field sparrows to the threat of nest predation. Auk 117:803–806. https://doi.org/10.1642/0004-8038(2000)117[0803:ATNROF]2.0.CO;2

Burhans DE (2001) Enemy recognition by field sparrows. Wilson Bull 113:189–193. https://doi.org/10.1676/0043-5643(2001)113[0189:ERBFS]2.0.CO;2

Butler PJ, Green JA, Boyd IL, Speakman JR (2004) Measuring metabolic rate in the field: the pros and cons of the doubly labelled water and heart rate methods. Funct Ecol 18:168–183. https://doi.org/10.1111/j.0269-8463.2004.00821.x

Byrkjedal I (1987) Antipredator behavior and breeding success in Greater Golden-Plover and Eurasian Dotterel. Condor 89:40–47. https://doi.org/10.2307/1368758

Campobello D, Sealy SG (2010) Enemy recognition of reed warblers (Acrocephalus scirpaceus): threats and reproductive value act independently in nest defence modulation. Ethology 116:498–508. https://doi.org/10.1111/j.1439-0310.2010.01764.x

Campobello D, Sealy SG (2018) Evolutionary significance of antiparasite, antipredator and learning phenotypes of avian nest defence. Sci Rep 8:10569. https://doi.org/10.1038/s41598-018-28275-3

Canoine V, Hayden TJ, Rowe K, Goymann W (2002) The stress response of European stonechats depends on the type of stressor. Behaviour 139:1303–1311. https://doi.org/10.1163/156853902321104172

Canty N, Gould JL (1995) The hawk/goose experiment: sources of variability. Anim Behav 50:1091–1095. https://doi.org/10.1016/0003-3472(95)80108-1

Carere C, Groothuis TGG, Mostl E, Daan S, Koolhaas JM (2003) Fecal corticosteroids in a territorial bird selected for different personalities: daily rhythm and the response to social stress. Horm Behav 43:540–548. https://doi.org/10.1016/S0018-506X(03)00065-5

Carlson NV, Pargeter HM, Templeton CN (2017a) Sparrowhawk movement, calling, and presence of dead conspecifics differentially impact blue tit (Cyanistes caeruleus) vocal and behavioral lobbing responses. Behav Ecol Sociobiol 71:133. https://doi.org/10.1007/s00265-017-2361-x

Carlson NV, Healy SD, Templeton CN (2017b) Hoo are you? Tits do not respond to novel predators as threats. Anim Behav 128:79–84. https://doi.org/10.1016/j.anbehav.2017.04.006

Caro T (2005) Antipredator defenses in birds and mammals. The University of Chicago Press, Chicago

Carrascal LM, Alonso CL (2006) Habitat use under latent predation risk. A case study with wintering forest birds. Oikos 112:51–62. https://doi.org/10.1111/j.0030-1299.2006.13787.x

Carrascal LM, Polo V (1999) Coal tits, Parus ater, lose weight in response to chases by predators. Anim Behav 58:281–285. https://doi.org/10.1006/anbe.1999.1142

Carter J, Lyons NJ, Cole HL, Goldsmith AR (2008) Subtle cues of predation risk: starlings respond to a predator's direction of eye-gaze. Proc Biol Sci 275:1709–1715. https://doi.org/10.1098/rspb.2008.0095

Cavanagh PM, Griffin CR (1993) Response of nesting common terns and laughing gulls to flyovers by large gulls. Wilson Bull 105:333–338

Cawthorn JM, Morris DL, Ketterson ED, Nolan V (1998) Influence of experimentally elevated testosterone on nest defence in Darkeyed Juncos. Anim Behav 56:617–621. https://doi.org/10.1006/anbe.1998.0849

Cerella J (1980) The pigeon's analysis of pictures. Pattern Recogn 12:1–6. https://doi.org/10.1016/0031-3203(80)90048-5

Cerella J (1986) Pigeons and perceptrons. Pattern Recogn 19:431–438. https://doi.org/10.1016/0031-3203(86)90041-5

Chandler CR, Rose RK (1988) Comparative analysis of the effects of visual and auditory stimuli on avian mobbing behaviour. J Field Ornithol 59:269–277

Chávez-Zichinelli CA, Gomez L, Ortiz-Pulido R, Lara C, Valdez R, Romano MC (2014) Testosterone levels in feces predict risk-sensitive foraging in hummingbirds. J Avian Biol 4:501–506. https://doi.org/10.1111/jav.00387

Chiver I, Jaramillo CA, Morton ES (2017) Mobbing behavior and fatal attacks on snakes by Fasciated Antshrikes (*Cymbilaimus lineatus*). J Ornithol 158:935–942. https://doi.org/10.1007/s10336-017-1452-9

Chu M (2001) Heterospecific responses to scream calls and vocal mimicry by phainopeplas (*Phainopepla nitens*) in distress. Behaviour 138:775–787. https://doi.org/10.1163/156853901752233406

Cibulková A, Veselý P, Fuchs R (2014) Importance of conspicuous colours in warning signals? The Great tit's (*Parus major*) point of view. Evol Ecol 28:427–439. https://doi.org/10.1007/s10682-014-9690-2

Clinchy M, Zanette L, Charlier TD, Newman AEM, Schmidt KL, Boonstra R, Soma KK (2011) Multiple measures elucidate glucocorticoid responses to environmental variation in predation threat. Oecologia 166:607–614. https://doi.org/10.1007/s00442-011-1915-2

Clode D, Birks JDS, Macdonald DW (2000) The influence of risk and vulnerability on predator mobbing by terns (*Sterna* spp.) and gulls (*Larus* spp.). J Zool 252:53–59. https://doi.org/10.1111/j.1469-7998.2000.tb00819.x

Cockrem JF (2013) Individual variation in glucocorticoid stress responses in animals. Gen Comp Endocr 181:45–58. https://doi.org/10.1016/j.ygcen.2012.11.025

Cockrem JF, Silverin B (2002) Sight of a predator can stimulate a corticosterone response in the great tit (*Parus major*). Gen Comp Endocr 125:248–255. https://doi.org/10.1006/gcen.2001.7749

Conkling TJ, Pope TL, Smith KN, Mathewson HA, Morrison ML, Wilkins RN, Cain JW (2012) Black-capped vireo nest predator assemblage and predictors for nest predation. J Wildlife Manag 76:1401–1411. https://doi.org/10.1002/jwmg.388

Conner LM, Rutledge JC, Smith LL (2010) Effects of mesopredators on nest survival of shrub-nesting songbirds. J Wildlife Manage 74:73–80. https://doi.org/10.2193/2008-406

Conover MR (1987) Acquisition of predator information by active and passive mobbers in ring-billed gull colonies. Behaviour 102:41–57. https://doi.org/10.2307/4534611

Conover MR, Perito JJ (1981) Response of starlings to distress calls and predator models holding conspecific prey. Z Tierpsychol 57:163–172. https://doi.org/10.1111/j.1439-0310.1981.tb01320.x

Cook NJ (2012) Review: Minimally invasive sampling media and the measurement of corticosteroids as biomarkers of stress in animals. Can J Anim Sci 92:227–259. https://doi.org/10.4141/cjas2012-045

Cook RG, Wright AA, Drachman EE (2013) Categorization of birds, mammals, and chimeras by pigeons. Behav Process 93:98–110. https://doi.org/10.1016/j.beproc.2012.11.006

Cordero PJ, Senar JC (1990) Interspecific nest defence in European Sparrows: different strategies to deal with a different species of opponent? Ornis Scan 21:71–73. https://doi.org/10.2307/3676381

Courter JR, Ritchison G (2010) Alarm calls of tufted titmice convey information about predator size and threat. Behav Ecol 21:936–942. https://doi.org/10.1093/beheco/arq086

Cox WA, Pruett MS, Benson TJ, Chiavacci SJ, Thompson FR III (2012) Development of camera technology for monitoring nests. USGS Northern Prairie Wildlife Research Center, Paper 250. https://doi.org/10.1525/california/9780520273139.003.0015

Cresswell W, Butler S, Whittingham MJ, Quinn JL (2009) Very short delays prior to escape from potential predators may function efficiently as adaptive risk-assessment periods. Behaviour 146:795–813. https://doi.org/10.1163/156853909X446217

Csermely D, Casagrande S, Calimero A (2006) Differential defensive response of common kestrels against a known or unknown predator. Ital J Zool 73:125–128. https://doi.org/10.1080/11250000600680072

Curio E (1975) The functional organization of anti-predator behaviour in the pied flycatcher: a study of avian visual perception. Anim Behav 23:1–115. https://doi.org/10.1016/0003-3472(75)90056-1

Curio E (1976) The ethology of predation. Springer, Berlin

Curio E, Onnebrink H (1995) Brood defense and brood size in the Great Tit (*Parus major*): a test of a model of unshared parental investment. Behav Ecol 6:235–241. https://doi.org/10.1093/beheco/6.3.235

Curio E, Regelmann K (1985) The behavioural dynamics of Great Tits (*Parus major*) approaching a predator. Z Tierpsychol 69:3–18. https://doi.org/10.1111/j.1439-0310.1985.tb00752.x

Curio E, Regelmann K (1986) Predator harassment implies a real deadly risk: a reply to Hennessy. Ethology 72:75–78. https://doi.org/10.1111/j.1439-0310.1986.tb00607.x

Curio E, Ernst U, Vieth W (1978) Cultural transmission of enemy recognition: one function of mobbing. Science 202:899–901. https://doi.org/10.1126/science.202.4370.899

Curio E, Klump G, Regelmann K (1983) An anti-predator response in the great tit (*Parus major*): is it tuned to predator risk? Oecologia 60:83–88. https://doi.org/10.1007/BF00379324

Cutler TL, Swann DE (1999) Using remote photography in wildlife ecology: a review. Wildlife Soc B 27:571–581

D'Orazio KA, Neudorf DLH (2008) Nest defense by Carolina Wrens. Wilson J Ornithol 120:467–472. https://doi.org/10.1676/06-149.1

da Cunha FCR, Fontenelle JCR, Griesser M (2017a) The presence of conspecific females influences male-mobbing behavior. Behav Ecol Sociobiol 71:52. https://doi.org/10.1007/s00265-017-2267-7

da Cunha FCR, Fontenelle JCR, Griesser M (2017b) Predation risk drives the expression of mobbing across bird species. Behav Ecol 28:1517–1523. https://doi.org/10.1093/beheco/arx111

Dale S, Gustavsen R, Slagsvold T (1996) Risk taking during parental care: a test of three hypotheses applied to the pied flycatcher. Behav Ecol Sociobiol 39:31–42. https://doi.org/10.1007/s002650050264

Davidson GL, Clayton NS, Thornton A (2015) Wild jackdaws, *Corvus monedula*, recognize individual humans and may respond to gaze direction with defensive behaviour. Anim Behav 108:17–24. https://doi.org/10.1016/j.anbehav.2015.07.010

Davies NB, Welbergen JA (2008) Cuckoo–hawk mimicry? An experimental test. Proc Biol Sci 275:1817–1822. https://doi.org/10.1098/rspb.2008.0331

Davies ZG, Fuller RA, Dallimer M, Loram A, Gaston KJ (2012) Household factors influencing participation in bird feeding activity: a national scale analysis. PLoS One 7:e39692. https://doi.org/10.1371/journal.pone.0039692

Deppe C, Holt D, Tewksbury J, Broberg L, Petersen J, Wood K (2003) Effect of northern pygmy-owl (*Glaucidium gnoma*) eyespots on avian mobbing. Auk 120:765–771. https://doi.org/10.2307/4090106

Dessborn L, Englund G, Elmberg J, Arzél C (2012) Innate responses of mallard ducklings towards aerial, aquatic and terrestrial predators. Behaviour 149:1299–1317. https://doi.org/10.1163/1568539X-00003014

Deviche P, Gao S, Davies S, Sharp PJ, Dawson A (2012) Rapid stress-induced inhibition of plasma testosterone in free-ranging male rufous-winged sparrows, *Peucaea carpalis*: characterization, time course, and recovery. Gen Comp Endocr 177:1–8. https://doi.org/10.1016/j.ygcen.2012.02.022

Dickens MJ, Bentley GE (2014) Stress, captivity, and reproduction in a wild bird species. Horm Behav 66:685–693. https://doi.org/10.1016/j.yhbeh.2014.09.011

Doherty J, Hoy R (1985) Communication in insects 3 – The auditory behavior of crickets – Some views of genetic coupling, song recognition, and predator detection. Q Rev Biol 60:457–472. https://doi.org/10.1086/414566

Duckworth JW (1991) Responses of breeding reed warblers *Acrocephalus scirpaceus* to mounts of sparrowhawk *Accipiter nisus*, cuckoo *Cuculus canorus* and jay *Garrulus glandarius*. Ibis 133:68–74. https://doi.org/10.1111/j.1474-919X.1991.tb04812.x

Duckworth JW (1997) Mobbing of a drongo cuckoo *Surniculus lugubris*. Ibis 139:190–192. https://doi.org/10.1111/j.1474-919X.1997.tb04526.x

Dufty AM, Crandall MB (2005) Corticosterone secretion in response to adult alarm calls in American Kestrels. J Field Ornithol 76:319–325. https://doi.org/10.1648/0273-8570-76.4.319

Dunn EH, Tessaglia DL (1994) Predation of birds at feeders in winter. J Field Ornithol 65:8–16

Duré Ruiz NM, Fasanella M, Fernández GJ (2018) Breeding southern house wrens exhibit a threat-sensitive response when exposed to different predator models. J Ethol 36:43–53. https://doi.org/10.1007/s10164-017-0528-6

Dutra LML, Young RJ, Galdinoa CAB, Vasconcellos AD (2016) Do apprehended saffron finches know how to survive predators? A careful look at reintroduction candidates. Behav Process 125:6–12. https://doi.org/10.1016/j.beproc.2016.01.007

East M (1981) Alarm calling and parental investment in the robin (*Erithacus rubecula*). Ibis 123:223–230. https://doi.org/10.1111/j.1474-919X.1981.tb00930.x

Edelaar P, Wright J (2006) Potential prey make excellent ornithologists: adaptive, flexible responses towards avian predation threat by Arabian Babblers *Turdoides squamiceps* living at a migratory hotspot. Ibis 148:664–671. https://doi.org/10.1111/j.1474-919X.2006.00567.x

Edwards G, Hosking E, Smith S (1950) Reactions of some passerine birds to a stuffed cuckoo. II. A detailed study of the Willow Warbler. Br Birds 43:144–150

Eggers S, Griesser M, Ekman J (2005) Predator-induced plasticity in nest visitation rates in the Siberian jay (*Perisoreus infaustus*). Behav Ecol 16:309–315. https://doi.org/10.1093/beheco/arh163

Ekman J, Eggers S, Griesser M, Tegelström H (2001) Queuing for preferred territories: delayed dispersal of Siberian Jays. J Anim Ecol 70:317–324. https://doi.org/10.1111/j.1365-2656.2001.00490.x

Elgar MA (1989) Predator vigilance and group size in mammals and birds: a critical review of the empirical evidence. Biol Rev 64:13–33. https://doi.org/10.1111/j.1469-185X.1989.tb00636.x

Elliot RD (1985) The effects of predation risk and group size on the anti-predator responses of nesting lapwings *Vanellus vanellus*. Behaviour 92:169–187. https://doi.org/10.1163/156853985X00433

Ely CR, Ward DH, Bollinger KS (1999) Behavioral correlates of heart rates of free-living Greater White-fronted Geese. Condor 101:390–395. https://doi.org/10.2307/1370002

Enstipp MR, Andrews RD, Jones DR (1999) Cardiac responses to first ever submergence in double-crested cormorant chicks (*Phalacrocorax auritus*). Comp Biochem Phys A 124:523–530. https://doi.org/10.1016/S1095-6433(99)00145-2

Evans CS, Evans L, Marler P (1993a) On the meaning of alarm calls: functional reference in an avian vocal system. Anim Behav 46:23–28. https://doi.org/10.1006/anbe.1993.1158

Evans C, Macedonia J, Marler P (1993b) Effects of apparent size and speed on the response of chickens, *Gallus gallus*, to computer-generated simulations of aerial predators. Anim Behav 46:1–11. https://doi.org/10.1006/anbe.1993.1156

Ficken MS, Ficken RW, Witkin SR (1978) Vocal repertoire of the black-capped chickadee. Auk 95:34–48. https://doi.org/10.2307/4085493

Ficken MS, Hailman ED, Hailman JP (1994) The chick-a-dee call system of the Mexican chickadee. Condor 96:70–82. https://doi.org/10.2307/1369065

Filiater TS, Breitwisch R, Nealen PM (1994) Predation on northern cardinals nests: does choice of nest site matter? Condor 96:761–768. https://doi.org/10.2307/1369479

Fisher RJ, Wiebe KL (2006) Investment in nest defense by Northern flickers: effects of age and sex. Wilson J Ornithol 118:452–460. https://doi.org/10.1676/05-117.1

Flasskamp A (1994) The adaptive significance of avian mobbing. An experimental test of the move on hypothesis. Ethology 96:322–333. https://doi.org/10.1111/j.1439-0310.1994.tb01020.x

Fluck E, Hogg S, Mabbutt PS, File SE (1996) Behavioural and neurochemical responses of male and female chicks to cat odour. Pharmacol Biochem Behav 54:85–91. https://doi.org/10.1016/0091-3057(95)02170-1

Folkers KL, Lowther PE (1985) Responses of nesting red-winged blackbirds and yellow warblers to brown-headed cowbirds. J Field Ornithol 56:175–177

Forbes M, Clark R, Weatherhead P, Armstrong T (1994) Risk-taking by female ducks: intra- and interspecific tests of nest defense theory. Behav Ecol Sociobiol 34:79–85. https://doi.org/10.1007/BF00164178

Frankenberg E (1981) The adaptive significance of avian mobbing IV. "Alerting others" and "perception advertisement" in blackbirds facing an owl. Z Tierpsychol 55:97–118

Fransson T, Weber TD (1997) Migratory fuelling in blackcaps (*Sylvia atricapilla*) under perceived risk of predation. Behav Ecol Sociobiol 41:75–80. https://doi.org/10.1007/s002650050366

Freeberg TM, Lucas JR (2002) Receivers respond differently to chick-a-dee calls varying in note composition in Carolina chickadees, *Poecile carolinensis*. Anim Behav 63:837–845. https://doi.org/10.1006/anbe.2001.1981

Friesen LE, Casbourn G, Martin V, Mackay RJ (2013) Nest predation in an anthropogenic landscape. Wilson J Ornithol 125:562–569. https://doi.org/10.1676/12-169.1

Fuchs E (1977) Predation and anti-predator behaviour in a mixed colony of terns *Sterna* sp. and Black-headed Gulls *Larus ridibundus* with special reference to the Sandwich Tern *Sterna sandvicensis*. Ornis Scand 8:17–32. https://doi.org/10.2307/3675984

Fuller RA, Warren PH, Armsworth PR, Barbosa O, Gaston KJ (2008) Garden bird feeding predicts the structure of urban avian assemblages. Divers Distrib 14:131–137. https://doi.org/10.1111/j.1472-4642.2007.00439.x

Gaddis P (1980) Mixed flocks, accipiters, and antipredator behavior. Condor 82:348–349. https://doi.org/10.2307/1367409

Gentle LK, Gosler AG (2001) Fat reserves and perceived predation risk in the great tit, *Parus major*. Proc Biol Sci 268:487–491. https://doi.org/10.1098/rspb.2000.1405

Gérard A, Jourdan H, Millon A, Vidal E (2015) Anti-predator behaviour in a procellariid seabird: wedge-tailed shearwaters do not respond to the odour of introduced ship rats: anti-predator behaviour in seabird. Austral Ecol 40:775–781. https://doi.org/10.1111/aec.12252

Ghalambor CK, Martin TE (2000) Parental investment strategies in two species of nuthatch vary with stage-specific predation risk and reproductive effort. Anim Behav 60:263–267. https://doi.org/10.1006/anbe.2000.1472

Gill SA, Sealy SG (1996) Nest defence by yellow warblers: recognition of a brood parasite and an avian nest predator. Behaviour 133:263–282. https://doi.org/10.1163/156853996X00143

Gill SA, Sealy SG (2004) Functional reference in an alarm signal given during nest defence: seet calls of yellow warblers denote brood-parasitic brown-headed cowbirds. Behav Ecol Sociobiol 56:71–80. https://doi.org/10.1007/s00265-003-0736-7

Gill SA, Neudorf DL, Sealy SG (1997a) Host responses to cowbirds near the nest: cues for recognition. Anim Behav 53:1287–1293. https://doi.org/10.1006/anbe.1996.0362

Gill SA, Grieef PM, Staib LM, Sealy SG (1997b) Does nest defence deter or facilitate cowbird parasitism? A test of the nesting-cue hypothesis. Ethology 103:56–71. https://doi.org/10.1111/j.1439-0310.1997.tb00007.x

Gilson WD, Kraitchman DL (2007) Cardiac magnetic resonance imaging in small rodents using clinical 1.5T and 3.0T scanners. Methods 43:35–45. https://doi.org/10.1016/j.ymeth.2007.03.012

Gluckman TL, Mundy NI (2013) Cuckoos in raptors' clothing: barred plumage illuminates a fundamental principle of Batesian mimicry. Anim Behav 86:1165–1181. https://doi.org/10.1016/j.anbehav.2013.09.020

Gochfeld M (1984) Antipredator behavior: aggressive and distraction displays of shorebirds. In: Burger J, Olla BL (eds) Behavior of marine animals 5. Plenum Press, New York

Godard RD, Bowers BB, Wilson CM (2007) Eastern bluebirds *Sialia sialis* do not avoid nest boxes with chemical cues from two common nest predators. J Avian Biol 38:128–131. https://doi.org/10.1111/j.2007.0908-8857.03788.x

Goethe F (1937) Beobachtungen und Untersuchungen zur Bioiogie der Silbermöwe (Larus a. argentatus Pontopp.) auf der Vogelinsel Memmertsand. J Ornithol 85:1–119. https://doi.org/10.1007/BF01905490

Goethe F (1940) Beobachtungen und Versuche über angeborene Schreckreaktionen junger Auerhühner. Z Tierpsychol 4:165–167. https://doi.org/10.1111/j.1439-0310.1940.tb00621.x

Goławski A, Meissner W (2007) The influence of territory characteristics and food supply on the breeding performance of the Red-backed Shrike (*Lanius collurio*) in an extensively farmed region of eastern Poland. Ecol Res 23:347–353. https://doi.org/10.1007/s11284-007-0383-y

Goławski A, Mitrus C (2008) What is more important: nest-site concealment or aggressive behaviour? A case study of the red-backed shrike, *Lanius collurio*. Folia Zool 57:403–410

Goodwin D (1953) The reactions of some nesting passerines towards live and stuffed Jays. Br Birds 46:193–200

Göth A (2001) Innate predator-recognition in Australian brush-turkey (*Alectura lathami*, Megapodiidae) hatchlings. Behaviour 138:117–136. https://doi.org/10.1163/156853901750077826

Gottfried BM (1979) Anti-predator aggression in birds nesting in old field habitats: an experimental analysis. Condor 81:251–257. https://doi.org/10.2307/1367626

Goymann W (2012) On the use of non-invasive hormone research in uncontrolled, natural environments: the problem with sex, diet, metabolic rate and the individual. Methods Ecol Evol 3:757–765. https://doi.org/10.1111/j.2041-210X.2012.00203.x

Green M, Green R, Carr WJ (1966) The hawk-goose phenomenon: a replication and an extension. Psychon Sci 4:185–186. https://doi.org/10.3758/BF03342241

Green R, Carr W, Green M (1968) The hawk:goose phenomenon: further confirmation and a search for the releaser. J Psychol 69:271–276. https://doi.org/10.1080/00223980.1968.10543474

Green RE, Hirons GJM, Kirby JS (1990) The effectiveness of nest defence by black-tailed godwits *Limosa limosa*. Ardea 78:405–413

Greig-Smith PW (1980) Parental investment in nest defence by stonechats (*Saxicola torquata*). Anim Behav 28:604–619. https://doi.org/10.1016/S0003-3472(80)80069-8

Griesser M (2003) Nepotistic vigilance behavior of Siberian jay parents. Behav Ecol 14:246–250. https://doi.org/10.1093/beheco/14.2.246

Griesser M (2008) Referential calls signal predator behavior in a group living bird species. Curr Biol 18:69–73. https://doi.org/10.1016/j.cub.2007.11.069

Griesser M (2009) Mobbing calls signal predator category in a kin group-living bird species. P Roy Soc Lond B Bio 276:2887–2892. https://doi.org/10.1098/rspb.2009.0551

Griesser M, Suzuki TN (2017) Naive juveniles are more likely to become breeders after witnessing predator mobbing. Am Nat 189:58–66. https://doi.org/10.1086/689477

Griffin AS, Savani RS, Hausmanis K, Lefebvre L (2005) Mixed-species aggregations in birds: Zenaida doves, *Zenaida aurita*, respond to the alarm calls of carib grackles, *Quiscalus lugubris*. Anim Behav 70:507–515. https://doi.org/10.1016/j.anbehav.2004.11.023

Griggio M, Matessi G, Pilastro A (2003) Male Rock Sparrow (*Petronia petronia*) nest defence correlates with female ornament size. Ethology 109:659–669. https://doi.org/10.1046/j.1439-0310.2003.00909.x

Griggio M, Fracasso G, Mahr K, Hoi H (2016) Olfactory assessment of competitors to the nest site: an experiment on a passerine species. PLoS One 11:e0167905. https://doi.org/10.1371/journal.pone.0167905

Grim T (2005) Host recognition of brood parasites: implications for methodology in studies of enemy recognition. Auk 122:530–543. https://doi.org/10.1642/0004-8038(2005)122[0530:HROBPI]2.0.CO;2

Grubb TC Jr (1977) Discrimination of aerial predators by American coots in nature. Anim Behav 25:1065–1066. https://doi.org/10.1016/0003-3472(77)90060-4

Guillory HD, LeBlanc DJ (1975) Mobbing and other interspecific aggression by Barn Swallows. Wilson Bull 87:110–112

Gunn JS, Desrochers A, Villard MA, Bourque J (2000) Playbacks of mobbing calls of black-capped chickadees as a method to estimate reproductive activity of forest birds. J Field Ornithol 71:472–483. https://doi.org/10.1648/0273-8570-71.3.472

Gyger M, Marler P, Pickert R (1987) Semantics of an avian alarm cal system: the male domestic fowl, *Gallus domesticus*. Behaviour 102:15–40. https://doi.org/10.1163/156853986X00027

Ha JM, Lee K, Yang EJ, Kim WJ, Song HK, Hwang IJ, Lee S, Jablonski PG (2018) Effect of nestlings' age on parentel responses to a predátory snake in *Parus minor*. Behaviour 155:327–336. https://doi.org/10.1163/1568539X-00003491

Haftorn S (2000) Contexts and possible functions of alarm calling in the willow tit, *Parus montanus*; the principle of 'better safe than sorry'. Behaviour 137:437–449. https://doi.org/10.1163/156853900502169

Hagelin JC, Jones IL (2007) Bird odors and other chemical substances: a defense mechanism or overlooked mode of intraspecific communication? Auk 124:741–761. https://doi.org/10.1642/0004-8038(2007)124[741:BOAOCS]2.0.CO;2

Hagelin JC, Jones IL, Rasmussen LEL (2003) A tangerine-scented social odour in a monogamous seabird. Proc Biol Sci 270:1323–1329. https://doi.org/10.1098/rspb.2003.2379

Hakkarainen H, Korpimäki E (1994) Nest defence of Tengmalm's owls reflects offspring survival prospects under fluctuating food conditions. Anim Behav 48:843–849. https://doi.org/10.1006/anbe.1994.1308

Hakkarainen H, Yli-Tuomi I, Korpimäki E, Ydenberg R (2002) Provisioning response to manipulation of apparent predation danger by parental pied flycatchers. Ornis Fenn 79:139–144

Halupka L (1999) Nest defence in an altricial bird with uniparental care: the influence of offspring age, brood size, stage of the breeding season and predator type. Ornis Fenn 76:97–105

Halupka K, Halupka L (1997) The influence of reproductive season stage on nest defence by meadow pipits (Anthus pratensis). Ethol Ecol Evol 9:89–98. https://doi.org/10.1080/08927014.1997.9522905

Hamer KC, Furness RW (1993) Parental investment and brood defense by male and female Great Skuas Catharacta skua: the influence of food-supply, laying date, body size and body condition. J Zool 230:7–18. https://doi.org/10.1111/j.1469-7998.1993.tb02668.x

Hamerstrom F (1957) The influence of a hawk's appetite on mobbing. Condor 59:192–194

Harkin A, O'Donnell JM, Kelly JP (2002) A study of VitalViewTM for behavioural and physiological monitoring in laboratory rats. Physiol Behav 77:65–77. https://doi.org/10.1016/S0031-9384(02)00810-7

Hartley P (1950) An experimental analysis of interspecific recognition. Sym Soc Exp Biol 4:313–336

Harvey PH, Greenwood PJ (1978) Anti-predator defence strategies: some evolutionary problems. In: Krebs JR, Davies NB (eds) Behavioural ecology: an evolutionary approach. Blackwell, Oxford

Hatch MI (1997) Variation in Song Sparrow nest defense: individual consistency and relationship to nest success. Condor 99:282–289. https://doi.org/10.2307/1369934

Hauser MD, Caffrey C (1994) Anti-predator response to raptor calls in wild crows, Corvus brachyrhynchos hesperis. Anim Behav 48:1469–1471. https://doi.org/10.1006/anbe.1994.1386

Hauser MD, Wrangham RW (1990) Recognition of predator and competitor calls in nonhuman primates and birds: a preliminary report. Ethology 86:116–130. https://doi.org/10.1111/j.1439-0310.1990.tb00423.x

Hendrichsen DK, Christiansen P, Nielsen EK, Dabelsteen T, Sunde P (2006) Exposure affects the risk of an owl being mobbed – experimental evidence. J Avian Biol 37:13–18

Herrnstein RJ, de Villiers PA (1980) Fish as natural category for people and pigeons. In: Bower GH (ed) The psychology of learning and motivation: advances in research and theory, vol 14. Academic Press, New York, pp 59–95

Herrnstein RJ, Loveland DH (1964) Complex visual concept in the pigeon. Science 146:549–551. https://doi.org/10.1126/science.146.3643.549

Herrnstein RJ, Loveland DH, Cable C (1976) Natural concepts in pigeons. J Exp Psych Anim Behav Process 2:285–311. https://doi.org/10.1037/0097-7403.2.4.285

Hill GE (1986) The function of distress calls given by tufted titmice (Parus bicolor). Anim Behav 34:590–598. https://doi.org/10.1016/S0003-3472(86)80128-2

Hilton GM, Cresswell W, Ruxton GD (1999) Intraflock variation in the speed of escape-flight response on attack by an avian predator. Behav Ecol 10:391–395. https://doi.org/10.1093/beheco/10.4.391

Hinde RA (1960) Factors governing the changes in strength of a partially inborn response as shown by the mobbing behaviour of the chaffinch (Fringilla coelebs). III. The interaction of short-term and long-term incremental and decremental effects. Proc Biol Sci 153:398–420. https://doi.org/10.1098/rspb.1961.0009

Hinsley SA, Bellamy PE, Moss D (1995) Sparrowhawk Accipiter nisus predation and feeding site selection by tits. Ibis 137:418–420. https://doi.org/10.1111/j.1474-919X.1995.tb08042.x

Hobson KA, Sealy SG (1989) Responses of yellow warblers to the threat of Cowbird parasitism. Anim Behav 38:510–519. https://doi.org/10.1016/S0003-3472(89)80044-2

Hobson KA, Bouchart ML, Sealy SG (1988) Reponses of naïve Yellow Warblers to a novel nest predator. Anim Behav 36:1823–1830. https://doi.org/10.1016/S0003-3472(88)80122-2

Hogstad O (1993) Nest defence and physical condition in Fieldfare *Turdus pilaris*. J Ornithol 134:25–33. https://doi.org/10.1007/BF01661130

Hogstad O (1995) Alarm calling by willow tits, *Parus montanus*, as mate investment. Anim Behav 49:221–223. https://doi.org/10.1016/0003-3472(95)80170-7

Hogstad O (2005) Sex-differences in nest defence in Fieldfares *Turdus pilaris* in relation to their size and physical condition. Ibis 147:375–380. https://doi.org/10.1111/j.1474-919x.2005.00416.x

Hollander FA, Van Overveld T, Tokka I, Matthysen E (2008) Personality and nest defence in the Great Tit (*Parus major*). Ethology 114:405–412. https://doi.org/10.1111/j.1439-0310.2008. 01488.x

Holway DA (1991) Nest-site selection and the importance of nest concealment in the black-throated blue warbler. Condor 93:575–581. https://doi.org/10.2307/1368189

Hölzer C, Bergmann H-H, McLean IG (1996) Training captive-raised, naive birds to recognise their predator. In: Ganslosser U, Hodges JK, Kaumanns W (eds) Research and captive propagation. Filander Verlag, Fürth

Honza M, Grim T, Čapek M Jr, Moksnes A, Røskaft E (2004) Nest defence, enemy recognition and nest inspection behaviour of experimentally parasitized reed warblers *Acrocephalus scirpaceus*. Bird Study 51:256–263. https://doi.org/10.1080/00063650409461361

Honza M, Šicha V, Procházka P, Ležalová R (2006) Host nest defense against a color-dimorphic brood parasite: great reed warblers (*Acrocephalus arundinaceus*) versus common cuckoos (*Cuculus canorus*). J Ornithol 147:629–637. https://doi.org/10.1007/s10336-006-0088-y

Huber L (2001) Visual categorization in pigeons. In: Cook RG (ed) Avian visual cognition. http:// www.pigeon.psy.tufts.edu/avc/

Huber L, Lenz R (1993) A test of the linear feature model of polymorphous concept discrimination with pigeons. Q J Exp Psychol 46B:1–18. https://doi.org/10.1080/14640749308401092

Jacobsen OW, Ugelvik M (1992) Antipredator behavior of breeding Eurasian wigeon. J Field Ornithol 63:324–330

Jansson C, Ekman J, von Brömssen A (1981) Winter mortality and food supply in Tits *Parus* spp. Oikos 37:313–322. https://doi.org/10.2307/3544122

Jenni-Eiermann S, Helfenstein F, Vallat A, Glauser G, Jenni L (2015) Corticosterone: effects on feather quality and deposition into feathers. Methods Ecol Evol 6:237–246. https://doi.org/10. 1111/2041-210X.12314

Jitsumori M (2004) Categorization and concept formation in pigeons: a perspective on comparative cognition. Int J Psychol 39(Suppl):373–373

Johnson-Delaney CA (2003) Use of the vetronics CardioStore for avian ECG. In: 24th Annual conference and expo of the Association-of-Avian-Veterinarians, Pittsburgh, PA, 26–28 Aug 2003. Take flight in Pittsburgh, Proceedings, pp 19–21

Johnstone CP, Reina RD, Lill A (2012) Interpreting indices of physiological stress in free-living vertebrates. J Comp Physiol B 182:861–879. https://doi.org/10.1007/s00360-012-0656-9

Kagawa H, Suzuki K, Takahasi M, Okanoya K (2014) Domestication changes innate constraints for birdsong learning. Behav Process 106:91–97. https://doi.org/10.1016/j.beproc.2014.04.011

Katz JS, Cook RG (2000) Stimulus repetition effects on texture-based visual search by pigeon. J Exp Psychol Anim Behav Process 26:220–236. https://doi.org/10.1037/0097-7403.26.2.220

Kelly DM, Cook RG (2003) Differential effects of visual context on pattern discrimination by pigeons (*Columba livia*) and humans (*Homo sapiens*). J Comp Psychol 117:200–208. https:// doi.org/10.1037/0735-7036.117.2.200

Kirkpatrick-Steger K, Wasserman EA, Biederman I (1996) Effects of spatial rearrangement of object components on picture recognition in pigeons. J Exp Anal Behav 65:465–475. https://doi. org/10.1901/jeab.1996.65-465

Kleindorfer S, Hoi H, Fessl B (1996) Alarm calls and chick reactions in the moustached warbler, *Acrocephalus melanopogon*. Anim Behav 51:1199–1206. https://doi.org/10.1006/anbe.1996.0125

Kleindorfer S, Fessl B, Hoi H (2003) The role of nest site cover for parental nest defence and fledging success in two *Acrocephalus warblers*. Avian Sci 3:21–29

Kleindorfer S, Fessl B, Hoi H (2005) Avian nest defence behaviour: assessment in relation to predator distance and type, and nest height. Anim Behav 69:307–313. https://doi.org/10.1016/j. anbehav.2004.06.003

Klump GM, Curio E (1983) Reactions of blue tits *Parus caeruleus* to hawk models of different sizes. Bird Behav 4:78–81

Knight RL, Temple SA (1986a) Nest defense in the *American goldfinch*. Anim Behav 34:879–897. https://doi.org/10.1016/S0003-3472(86)80075-6

Knight RL, Temple SA (1986b) Methodological problems in studies of avian nest defence. Anim Behav 34:561–566. https://doi.org/10.1016/S0003-3472(86)80125-7

Knight RL, Temple SA (1986c) Why does intensity of avian nest defence increase during the nesting cycle? Auk 103:318–327

Knight RL, Temple SA (1988) Nest-defense behavior in the red-winged blackbird. Condor 90:193–200. https://doi.org/10.2307/1368448

Koivula K, Rytkönen S, Orell M (1995) Hunger-dependency of hiding behaviour after a predator attack in dominant and subordinate willow tits. Ardea 83:397–404

Krams I (2000) Length of feeding day and body weight of great tits in a single and a two-predator environment. Behav Ecol Sociobiol 48:147–153. https://doi.org/10.1007/s002650000214

Krams I (2001) Communication in crested tits and the risk of predation. Anim Behav 61:1065–1068. https://doi.org/10.1006/anbe.2001.1702

Krams I, Krama T, Igaune K (2006) Alarm calls of wintering great tits *Parus major*: warning of mate, reciprocal altruism or a message to the predator? J Avian Biol 37:131–136. https://doi.org/ 10.1111/j.0908-8857.2006.03632.x

Krätzig H (1940) Untersuchungen zur Lebensweise des Moorschneehuhns (*Lagopus l. lagopus* L.) während der Jugendentwicklung. J Ornithol 88:139–165. https://doi.org/10.1007/BF01670363

Królikowska N, Szymkowiak J, Laidlaw RA, Kuczyński L (2016) Threat-sensitive anti-predator defence in precocial wader, the northern lapwing *Vanellus vanellus*. Acta Ethol 19:163–171. https://doi.org/10.1007/s10211-016-0236-1

Kruuk H (1964) Predators and anti-predator behaviour of the black-headed gull (*Larus ridibundus*). Behav Suppl 11:1–129. https://doi.org/10.2307/4511169

Kruuk H (1976) The biological function of gulls' attraction towards predators. Anim Behav 24:146–153. https://doi.org/10.1016/S0003-3472(76)80108-X

Kullberg C (1998) Spatial niche dynamics under predation risk in the Willow Tit *Parus montanus*. J Avian Biol 29:235–240. https://doi.org/10.2307/3677105

Kullberg C, Lind J (2002) An experimental study of predator recognition in great tit fledglings. Ethology 108:429–441. https://doi.org/10.1046/j.1439-0310.2002.00786.x

Kumar A (2003) Acoustic communication in birds. Resonance 8:44–55. https://doi.org/10.1007/ BF02837868

Lack D (1943) The life of the robin. Whiterby, London

Lack D (1954) The natural regulation of animal numbers. Clarendon Press, Oxford

Lane J (2006) Can non-invasive glucocorticoid measures be used as reliable indicators of stress in animals? Anim Welfare 15:331–342

Langmore NE, Feeney WE, Crowe-Riddell J, Luan H, Louwrens KM, Cockburn A (2012) Learned recognition of brood parasitic cuckoos in the superb fairy-wren *Malurus cyaneus*. Behav Ecol 23:798–805. https://doi.org/10.1093/beheco/ars033

Lazareva OF, Freiburger KL, Wasserman EA (2004) Pigeons concurrently categorize photographs at both basic and superordinate levels. Psychon B Rev 11:1111–1117. https://doi.org/10.3758/ BF03196745

Lazareva OF, Freiburger KL, Wasserman EA (2006) Effects of stimulus manipulations on visual categorization in pigeons. Behav Process 72:224–233. https://doi.org/10.1016/j.beproc.2006.03.004

Leger DW, Carroll LF (1981) Mobbing calls of the Phainopepla. Condor 83:377–380. https://doi. org/10.2307/1367509

Lemmetyinen R (1971) Nest defence behaviour of Common and Arctic Terns and its effects on the success achieved by predators. Ornis Fenn 48:13–24

Liang W, Møller AP (2015) Hawk mimicry in cuckoos and anti-parasitic aggressive behavior of barn swallows in Denmark and China. J Avian Biol 46:216–223. https://doi.org/10.1111/jav.00515

Lilliendahl K (1997) The effect of predator presence on body mass in captive greenfinches. Anim Behav 53:75–81. https://doi.org/10.1006/anbe.1996.0279

Lima SL (1993) Ecological and evolutionary perspectives on escape from predatory attack: a survey of North American birds. Wilson Bull 105:1–47

Lima SL (1995) Back to the basics of anti-predatory vigilance: the group size effect. Anim Behav 49:11–20. https://doi.org/10.1016/0003-3472(95)80149-9

Lima SL, Dill LM (1990) Behavioral decisions made under the risk of predation – a review and prospectus. Can J Zool 68:619–640. https://doi.org/10.1139/z90-092

Lind J, Jongren F, Nilsson J, Alm DS, Strandmark A (2005) Information, predation risk and foraging decisions during mobbing in great tits *Parus major*. Ornis Fenn 82:89–96

Londoño GA, García DA, Sánchez Martínez MA (2015) Morphological and behavioral evidence of Batesian mimicry in nestlings of a lowland Amazonian bird. Am Nat 185:135–141. https://doi.org/10.1016/0003-3472(95)80149-9

Lorenz K (1937a) The companion in the bird's world. Auk 54:245–273. https://doi.org/10.2307/4078077

Lorenz K (1937b) Über die Bildung des Instinktbegriffes. Naturwissenschaften 25:289–300. https://doi.org/10.1007/BF01492648

Lorenz K (1939) Vergleichende verhaltensforschung. Verh Deut Z 12:69–102

Lubow RE (1974) High-order concept formation in the pigeon. J Exp Anal Behav 21:475–483. https://doi.org/10.1901/jeab.1974.21-475

Lyon BE, Gilbert GS (2013) Rarely parasitized and unparasitized species mob and alarm call to cuckoos: implications for sparrowhawk mimicry by brood parasitic cuckoos. Wilson J Ornithol 125:627–630. https://doi.org/10.1676/12-162.1

Magrath RD, Pitcher BJ, Gardner JL (2007) A mutual understanding? Interspecific responses by birds to each other's aerial alarm calls. Behav Ecol 18:944–951. https://doi.org/10.1093/beheco/arm063

Magrath RD, Haff TM, Horn AG, Leonard ML (2010) Calling in the face of danger: predation risk and acoustic communication by parent birds and their offspring. Adv Stud Behav 41:187–253. https://doi.org/10.1016/S0065-3454(10)41006-2

Maloney RF, McLean IG (1995) Historical and experimental learned predator recognition in free-living New-Zealand robins. Anim Behav 50:1193–1201. https://doi.org/10.1016/0003-3472(95)80036-0

Mark D, Stutchbury BJ (1994) Response of a forest-interior songbird to the threat of cowbird parasitism. Anim Behav 47:275–280. https://doi.org/10.1006/anbe.1994.1039

Marr D, Nishihara HK (1978) Representation and recognition of spatial-organization of 3-dimensional shapes. Proc Biol Sci 200:269–294. https://doi.org/10.1098/rspb.1978.0020

Martin TE (1993a) Nest predation among vegetation layers and habitat types: revising the dogmas. Am Nat 141:897–913. https://doi.org/10.1086/285515

Martin TE (1993b) Nest predation and nest sites. Bioscience 43:523–532. https://doi.org/10.2307/1311947

Martin TE, Roper JJ (1988) Nest predation and nest-site selection of a western population of the hermit thrush. Condor 90:51–57. https://doi.org/10.2307/1368432

Martin J, de Neve L, Polo V, Fargallo JA, Soler M (2006) Health-dependent vulnerability to predation affects escape responses of unguarded chinstrap penguin chicks. Behav Ecol Sociobiol 60:778–784. https://doi.org/10.1007/s00265-006-0221-1

Mathot KJ, van den Hout PJ, Piersma T (2009) Differential responses of red knots, *Calidris canutus*, to perching and flying sparrowhawk, *Accipiter nisus*, models. Anim Behav 77:1179–1185. https://doi.org/10.1016/j.anbehav.2009.01.024

Matsukawa A, Inoue S, Jitsumori M (2004) Pigeon's recognition of cartoons: effects of fragmentation, scrambling, and deletion of elements. Behav Process 65:25–34. https://doi.org/10.1016/S0376-6357(03)00147-5

Maziarz M, Piggott C, Burgess M (2018) Predator recognition and differential behavioural responses of adult wood warblers *Phylloscopus sibilatrix*. Acta Ethol 21:13–20. https://doi.org/10.1007/s10211-017-0275-2

McIvor GE, Lee VE, Thornton A (2018) Testing social leasing of anti-predator responses in juvenilie jackdaws: the importace of accounting for levels of agitation. R Soc Open Sci 5:171571. https://doi.org/10.1098/rsos.171571

McLean IG (1987) Response to a dangerous enemy: should a brood parasite be mobbed? Ethology 75:235–245. https://doi.org/10.1111/j.1439-0310.1987.tb00656.x

McLean IG, Rhodes G (1991) Enemy recognition and response in birds. Curr Ornithol 8:173–211

McLean IG, Smith JNM, Stewart KG (1986) Mobbing behaviour, nest exposure, and breeding success in the American robin. Behaviour 96:171–185. https://doi.org/10.1163/156853986X00270

McLean IG, Lundie-Jenkins G, Jarman PJ (1996) Teaching an endangered mammal to recognise predators. Biol Conserv 75:51–62. https://doi.org/10.1016/0006-3207(95)00038-0

McLean IG, Hölzer C, Studholme BJ (1999) Teaching predator-recognition to a naive bird: implications for management. Biol Conserv 87:123–130. https://doi.org/10.1016/S0006-3207(98)00024-X

McNicholl MK (1973) Habituation of aggressive responses to avian predators by terns. Auk 90:902–904. https://doi.org/10.2307/4084379

McPhail LT, Jones DR (1998) The relationship between power output and heart rate in ducks diving voluntarily. Comp Biochem Phys A 120:219–225. https://doi.org/10.1016/S1095-6433(98)00005-1

Meilvang D, Moksnes A, Røskaft E (1997) Nest predation, nesting characteristics and nest defence behaviour of Fieldfares and Redwings. J Avian Biol 28:331–337. https://doi.org/10.2307/3676947

Melvin KB, Cloar FT (1969) Habituation of responses of quail (*Colinus virginianus*) to a Hawk (*Buteo swainsoni*): measurement through an 'innate suppression' technique. Anim Behav 17:468–473. https://doi.org/10.1016/0003-3472(69)90148-1

Melzack R, Penick E, Beckett A (1959) The problem of innate fear of the hawk shape: an experimental study with mallard ducks. J Camp Physiol Psychol 52:69–698. https://doi.org/10.1037/h0038532

Michl G, Török J, Garamszegi LZS, Tóth L (2000) Sex-dependent risk taking in the collared flycatcher, *Ficedula albicollis*, when exposed to a predator at the nestling stage. Anim Behav 59:623–628. https://doi.org/10.1006/anbe.1999.1352

Miller L (1952) Auditory recognition of predators. Condor 54:89–92. https://doi.org/10.2307/1364595

Minderman J, Lind J, Cresswell W (2006) Behaviourally mediated indirect effects: interference competition increases predation mortality in foraging redshanks. J Anim Ecol 75:713–723. https://doi.org/10.1111/j.1365-2656.2006.01092.x

Moksnes A, Røskaft E, Braa AT, Korsnes L, Lampe HT, Pedersen HC (1990) Behavioural responses of potential hosts towards artificial Cuckoo eggs and dummies. Behaviour 116:64–89. https://doi.org/10.1163/156853990X00365

Montgomerie RD, Weatherhead PJ (1988) Risks and rewards of nest defence by parent birds. Q Rev Biol 63:167–187. https://doi.org/10.1086/415838

Moore EL, Mueller HC (1982) Cardiac responses of domestic chickens to hawk and goose models. Behav Proc 7:255–258. https://doi.org/10.1016/0376-6357(82)90040-7

Mueller HC, Parker PG (1980) Naive ducklings show different cardiac response to hawk than to goose models. Behaviour 74:101–113. https://doi.org/10.1163/156853980X00339

Müller M, Pasinelli G, Schiegg K, Spaar R, Jenni L (2005) Ecological and social effects on reproduction and local recruitment in the red-backed shrike. Oecologia 143:37–50. https://doi.org/10.1007/s00442-004-1770-5

Murphy TG (2006) Predator-elicited visual signal: why the turquoise-browed motmot wag-displays its racketed tail. Behav Ecol 17:547–553. https://doi.org/10.1093/beheco/arj064

Murray L (2015) Success and predation of bird nests in grasslands at Valley Forge National Historical Park. Northeast Nat 22:10–19. https://doi.org/10.1656/045.022.0103

Nácarová J, Veselý P, Fuchs R (2018) Effect of the exploratory behaviour on a bird's ability to categorize a predator. Behav Process 151:89–95. https://doi.org/10.1016/j.beproc.2018.03.021

Naguib M, Mundry R, Ostreiher R, Hultsch H, Schrader L, Todt D (1999) Cooperatively breeding Arabian babblers call differently when mobbing in different predator-induced situations. Behav Ecol 10:636–640. https://doi.org/10.1093/beheco/10.6.636

Naguib M, Janik V, Clayton N, Zuberbühler K (2009) Vocal communication in birds and mammals. Academic Press

Nealen PM, Breitwisch R (1997) Northern cardinal sexes defend nests equally. Wilson Bull 109:269–278

Němec M, Fuchs R (2014) Nest defense of the red-backed shrike Lanius collurio against five corvid species. Acta Ethol 17:149–154. https://doi.org/10.1007/s10211-013-0175-z

Němec M, Syrová M, Dokoupilová L, Veselý P, Šmilauer P, Landová E, Lišková S, Fuchs R (2015) Surface texture plays important role in predator recognition by Red-backed Shrikes in field experiment. Anim Cogn 18:259–268. https://doi.org/10.1007/s10071-014-0796-2

Němec M, Součková T, Fuchs R (unpublished) The red-backed shrike Lanius collurio recognize the predator due local, not due global features

Neudorf DL, Sealy SG (1992) Reactions of four passerine species to threats of predation and cowbird parasitism: enemy recognition or generalized responses? Behaviour 123:84–105. https://doi.org/10.1163/156853992X00138

Neudorf DLH, Sears KE, Sealy SG (2011) Responses of nesting yellow-headed blackbirds and yellow warblers to wrens. Wilson J Ornithol 123:823–827. https://doi.org/10.2307/41480554

Neumann PG (1977) Visual prototype formation with discontinuous representation of dimensions of variability. Mem Cogn 5:187–197. https://doi.org/10.3758/BF03197361

Nice MM, Ter Pelkwyk J (1941) Enemy recognition by the song sparrow. Auk 58:195–214. https://doi.org/10.2307/4079104

Nicholls E, Ryan CME, Bryant CML, Lea SEG (2011) Labeling and family resemblance in the discrimination of polymorphous categories by pigeons. Anim Cogn 14:21–34. https://doi.org/10.1007/s10071-010-0339-4

Nijman V (2004) Seasonal variation in naturally occurring mobbing behaviour of drongos (Dicruridae) towards two avian predators. Ethol Ecol Evol 16:25–32. https://doi.org/10.1080/08927014.2004.9522651

Nilsson SG (1984) The evolution of nest site selection among hole-nesting birds: the importance of nest predation and competition. Ornis Scand 15:167–175. https://doi.org/10.2307/3675958

Nocera JJ, Ratcliffe LM (2009) Migrant and resident birds adjust antipredator behavior in response to social information accuracy. Behav Ecol 21:121–128. https://doi.org/10.1093/beheco/arp161

Nováková N, Veselý P, Fuchs R (2017) Object categorization by wild ranging birds – winter feeder experiments. Behav Process 143:7–12. https://doi.org/10.1016/j.beproc.2017.08.002

Obozova TA, Smirnova AA, Zorina ZA (2012) Relational learning in glaucous-winged gulls (Larus glaucescens). Span J Psychol 15:873–880. https://doi.org/10.5209/rev_SJOP.2012.v15.n3.39380

Olendorf R, Robinson SK (2000) Effectiveness of nest defence in the Acadian Flycatcher Empidonax virescens. Ibis 142:365–371. https://doi.org/10.1111/j.1474-919X.2000.tb04432.x

Onnebrink H, Curio E (1991) Brood defence and age of young: a test of the vulnerability hypothesis. Behav Ecol Sociobiol 29:61–68. https://doi.org/10.1007/BF00164296

Orchinik M (1998) Glucocorticoids, stress, and behavior: shifting the timeframe. Horm Behav 34:320–327. https://doi.org/10.1006/hbeh.1998.1488

Overskaug K, Sunde P, Stuve G (2000) Intersexual differences in the diet composition of Norwegian raptors. Ornis Norvegica 23:24–30

Owen RB (1969) Heart rate, a measure of metabolism in the blue-winged teal. Comp Biochem Physiol 31:431–436. https://doi.org/10.1016/0010-406X(69)90024-3

Palestis BG (2005) Nesting stage and nest defense by common terns. Waterbirds 28:87–94. https://doi.org/10.1675/1524-4695(2005)028[0087:NSANDB]2.0.CO;2

Palleroni A, Hauser M, Marler P (2005) Do responses of galliform birds vary adaptively with predator size? Anim Cogn 8:200–210. https://doi.org/10.1007/s10071-004-0250-y

Patterson TL, Petrinovich L, James DK (1980) Reproductive value and appropriateness of response to predators by white-crowned sparrows. Behav Ecol Sociobiol 7:227–231. https://doi.org/10.1007/BF00299368

Pavel V, Bureš S (2001) Offspring age and nest defence: test of the feedback hypothesis in the meadow pipit. Anim Behav 61:297–303. https://doi.org/10.1006/anbe.2000.1574

Pearce JM (1994) Discrimination and categorization. Anim Learn Cogn 9:109–134

Pearce JM (2008) Animal learning and cognition, an introduction, 3rd edn. Psychology Press, Hove

Peluc SI, Sillett TS, Rotenberry JT, Ghalambor CK (2008) Adaptive phenotypic plasticity in an island songbird exposed to a novel predation risk. Behav Ecol 19:830–835. https://doi.org/10.1093/beheco/arn033

Poiani A, Yorke M (1989) Predator harassment: more evidence on the deadly risk. Ethology 83:167–169. https://doi.org/10.1111/j.1439-0310.1989.tb00526.x

Posner MI, Keele SW (1968) On the genesis of abstract ideas. J Exp Psychol 77:353–363. https://doi.org/10.1037/h0025953

Pravosudov VV, Grubb TCJ (1998) Management of fat reserves in tufted titmice, *Baelophus bicolor*, in relation to risk of predation. Anim Behav 56:49–54. https://doi.org/10.1006/anbe.1998.0739

Quinn JL, Cresswell W (2005) Personality, anti-predation behaviour and behavioural plasticity in the chaffinch *Fringilla coelebs*. Behaviour 142:1383–1408. https://doi.org/10.1163/156853905774539391

Quinn JL, Cole EF, Bates J, Payne RW, Cresswell W (2012) Personality predicts individual responsiveness to the risks of starvation and predation. Proc Biol Sci 279:1919–1926. https://doi.org/10.1098/rspb.2011.2227

Radford AN, Blakey JK (2000) Intensity of nest defence is related to offspring sex ratio in the Great Tit *Parus major*. Proc Biol Sci 267:535–538. https://doi.org/10.1098/rspb.2000.1033

Rainey HJ, Zuberbuhler K, Slater PJB (2004) The responses of Black-casqued Hornbills to predator vocalisations and primate alarm calls. Behaviour 141:1263–1277. https://doi.org/10.1163/1568539042729658

Randler C (2006) Disturbances by dog barking increase vigilance in coots *Fulica atra*. Eur J Wildl Res 52:265–270. https://doi.org/10.1007/s10344-006-0049-z

Rasa OAE (1981) Raptor recognition: an interspecific tradition? Naturwissenschaften 68:151–152. https://doi.org/10.1007/BF01047264

Redondo T (1989) Avian nest defence: theoretical models and evidence. Behaviour 111:161–195. https://doi.org/10.2307/4534813

Redondo T, Carranza J (1989) Offspring reproductive value and nest defense in the magpie (*Pica pica*). Behav Ecol Sociobiol 25:369–378. https://doi.org/10.1007/BF00302995

Remeš V (2005) Nest concealment and parental behaviour interact in affecting nest survival in the blackcap (*Sylvia atricapilla*): an experimental evaluation of the parental compensation hypothesis. Behav Ecol Sociobiol 58:326–332. https://doi.org/10.1007/s00265-005-0910-1

Reudink MW, Nocera JJ, Curry RL (2007) Anti-predator responses of Neotropical resident and migrant birds to familiar and unfamiliar owl vocalizations on the Yucatan peninsula. Ornitol Neotrop 18:543–552

Ricklefs RE (1969) An analysis of nesting mortality in birds. Smithson Contrib Zool 9:1–48

Robb GN, McDonald RA, Chamberlain DE, Reynolds SJ, Harrison THE, Bearhop S (2008a) Winter feeding of birds increases productivity in the subsequent breeding season. Biol Lett 4:220–223. https://doi.org/10.1098/rsbl.2007.0622

Robb GN, McDonald RA, Chamberlain DE, Reynolds SJ, Bearhop S (2008b) Food for thought: supplementary feeding as a driver of ecological change in avian populations. Front Ecol Environ 6:476–484. https://doi.org/10.1890/060152

Robertson RJ, Norman RF (1977) The function and evolution of aggressive host behaviour toward the Brown-Headed Cowbird (*Molothrus ater*). Can J Zool 55:508–518. https://doi.org/10.1139/z77-066

Rodriguez-Prieto I, Fernández-Juricic E, Martín J, Regis Y (2009) Antipredator behavior in black-birds: habituation complements risk allocation. Behav Ecol 20:371–377. https://doi.org/10.1093/beheco/arn151

Roos S, Pärt T (2004) Nest predators affect spatial dynamics of breeding red-backed shrikes (*Lanius collurio*). J Anim Ecol 73:117–127. https://doi.org/10.1111/j.1365-2656.2004.00786.x

Roth TC, Cox JG, Lima SL (2008) The use and transfer of information about predation risk in flocks of wintering finches. Ethology 114:1218–1226. https://doi.org/10.1111/j.1439-0310.2008.01572.x

Rothstein SI (1990) A model system for coevolution: avian brood parasitism. Annu Rev Ecol Syst 21:481–508. https://doi.org/10.1146/annurev.es.21.110190.002405

Rytkönen S (2002) Nest defence in great tits *Parus major*: support for parental investment theory. Behav Ecol Sociobiol 52:379–384. https://doi.org/10.1007/s00265-002-0530-y

Rytkönen S, Koivula K, Orell M (1990) Temporal increase in nest defence intensity of the willow tit (*Parus montanus*); parental investment or methodological artefact? Behav Ecol Sociobiol 27:283–286. https://doi.org/10.1007/BF00164901

Rytkönen S, Orell M, Koivula K, Soppela M (1995) Correlation between two components of parental investment: nest defence intensity and nestling provisioning effort of willow tits. Oecologia 104:386–393. https://doi.org/10.1007/BF00328375

Sansom A, Lind J, Cresswell W (2009) Individual behavior and survival: the roles of predator avoidance, foraging success, and vigilance. Behav Ecol 20:1168–1174. https://doi.org/10.1093/beheco/arp110

Scaife M (1976) The response to eye-like shapes by birds. I. The effect of context: a predator and a strange bird. Anim Behav 24:195–199. https://doi.org/10.1016/S0003-3472(76)80115-7

Schaller GB, Emlen JT Jr (1962) The ontogeny of avoidance behaviour in some precocial birds. Anim Behav 10:370–381. https://doi.org/10.1016/0003-3472(62)90060-X

Schetini de Azevedo C, Young RJ, Rodrigues M (2012) Failure of captive-born greater rheas (*Rhea americana*, Rheidae, Aves) to discriminate between predator and nonpredator models. Acta Ethol 15:179–185. https://doi.org/10.1007/s10211-012-0124-2

Scheuerlein A, Van't Hof TJ, Gwinner E (2001) Predators as stressors? Physiological and reproductive consequences of predation risk in tropical stonechats (*Saxicola torquata* axillaris). Proc Biol Sci 268:1575–1582. https://doi.org/10.1098/rspb.2001.1691

Schleidt W, Shalter MD, Moura-Neto H (2011) The hawk/goose story: the classical ethological experiments of Lorenz and Tinbergen, revisited. J Comp Psychol 125:121–133. https://doi.org/10.1037/a0022068

Shalter MD (1978) Mobbing in the Pied Flycatcher: effect of experiencing a live owl on responses to a stuffed facsimile. Z Tierpsychol 47:173–179. https://doi.org/10.1111/j.1439-0310.1978.tb01828.x

Shanks DR (1994) Human associative learning. In: Mackintosh NJ (ed) Animal learning and cognition. Academic Press, San Diego, CA, pp 335–368

Shedd DH (1982) Seasonal variation and function of mobbing and related antipredatory behaviors of the American robin (*Turdus migratorius*). Auk 99:342–346

Shedd DH (1983) Seasonal variation in mobbing intensity in the Black-capped chickadee. Wilson Bull 95:343–348

Shettleworth SJ (2010) Cognition, evolution and behaviour, 2nd edn. Oxford University Press, Oxford

Shields WM (1984) Barn swallow mobbing: self-defence, collateral kin defence, group defence, or parental care? Anim Behav 32:132–148. https://doi.org/10.1016/S0003-3472(84)80331-0

Siegel-Causey D, Hunt GL Jr (1981) Colonial defense behavior in double-crested and pelagic cormorants. Auk 98:522–531

Sieving KE, Hetrick SA, Avery ML (2010) The versatility of graded acoustic measures in classification of predation threats by the tufted titmouse *Baeolophus bicolor*: exploring a mixed framework for threat communication. Oikos 119:264–276. https://doi.org/10.1111/j.1600-0706.2009.17682.x

Simmons KEL (1955) The nature of the predator reactions of waders towards humans, with special reference to the role of the aggressive-, escape and brooding-drives. Behaviour 8:130–173. https://doi.org/10.1163/156853955X00201

Skinner BF (1931) The concept of the reflex in the description of behavior. J Gen Psychol 5:427–458. https://doi.org/10.1080/00221309.1931.9918416

Skutch AF (1949) Do tropical birds rear as many young as they can nourish? Ibis 91:430–458. https://doi.org/10.1111/j.1474-919X.1949.tb02293.x

Skutch AF (1996) Antbirds and ovenbirds: their lives and homes. University of Texas Press, Austin

Smirnova AA, Bagotskaya MS, Zorina ZA (2003) A larger set of training stimuli does not facilitate matching learning in crows. Zh Vyssh Nerv Deyat 53:321–328

Smith EL (1970) Cactus Wrens attack ground squirrel. Condor 72:363–364. https://doi.org/10.2307/1366016

Smith SM (1997) The black-capped chickadee. Behavioral ecology and natural history. Cornell University Press, Ithaca

Smith JM, Graves HB (1978) Some factors influencing mobbing behaviour in barn swallows *Hirundo rustica*. Behav Biol 23:355–372. https://doi.org/10.1016/S0091-6773(78)91379-2

Smith JNM, Arcese P, McLean IG (1984) Age, experience, and enemy recognition by wild song sparrows. Behav Ecol Sociobiol 14:101–106. https://doi.org/10.1007/BF00291901

Soard CM, Ritchison G (2009) "Chick-a-dee" calls of Carolina chickadees convey information about degree of threat posed by avian predators. Anim Behav 78:1447–1453. https://doi.org/10.1016/j.anbehav.2009.09.026

Sordahl TA (1990) The risks of avian mobbing and distraction behavior: an anecdotal review. Wilson Bull 102:349–352

Sordahl TA (2004) Field evidence of predator discrimination abilities in American Avocets and Black-necked stilts. J Field Ornithol 75:376–385. https://doi.org/10.1648/0273-8570-75.4.376

Stenhouse IJ, Gilchrist HG, Montevecchi WA (2005) An experimental study examining the anti-predator behaviour of Sabine's gulls (*Xema sabini*) during breeding. J Ethol 23:103–108. https://doi.org/10.1007/s10164-004-0135-1

Stoddard MC (2012) Mimicry and masquerade from the avian visual perspective. Curr Zool 58:630–648. https://doi.org/10.1093/czoolo/58.4.630

Stone E, Trost CH (1991) Predators, risks and context for mobbing and alarm calls in black-billed magpies. Anim Behav 41:633–638. https://doi.org/10.1016/S0003-3472(05)80901-7

Storch S, Gremillet D, Culik BM (1999) The telltale heart: a non-invasive method to determine the energy expenditure of incubating great cormorants *Phalacrocorax carbo* carbo. Ardea 87:207–215

Strnad M, Němec M, Veselý P, Fuchs R (2012) Red-backed Shrikes (*Lanius collurio*) adjust the mobbing intensity, but not mobbing frequency, by assessing the potential threat to themselves from different predators. Ornis Fennica 89:206–215

Strnadová I, Němec M, Strnad M, Veselý P, Fuchs R (2018) The nest defence by the red-backed shrike *Lanius collurio* – support for the vulnerability hypothesis. J Avian Biol 49:e01726. https://doi.org/10.1111/jav.01726

Suková K, Uchytilová M, Lindová J (2013) Abstract concept formation in African grey parrots (*Psittacus erithacus*) on the basis of a low number of cues. Behav Process 96:36–41. https://doi.org/10.1016/j.beproc.2013.02.008

Suzuki TN (2011) Parental alarm calls warn nestlings about different predatory threats. Curr Biol 21:R15–R16. https://doi.org/10.1016/j.cub.2010.11.027

Suzuki TN (2012) Referential mobbing calls elicit different predator-searching behaviours in Japanese great tits. Anim Behav 84:53–57. https://doi.org/10.1016/j.anbehav.2012.03.030

Syrová M, Němec M, Veselý P, Landová E, Fuchs R (2016) Facing a clever predator demands clever responses – Red-Backed Shrikes (*Lanius collurio*) vs. Eurasian Magpie (*Pica pica*). PLoS One 11:e0159432. https://doi.org/10.1371/journal.pone.0159432

Templeton CN, Greene E (2007) Nuthatches eavesdrop on variations in heterospecific chickadee mobbing alarm calls. Proc Natl Acad Sci USA 104:5479–5482. https://doi.org/10.1073/pnas.0605183104

Templeton CN, Greene E, Davis K (2005) Allometry of alarm calls: black-capped chickadees encode information about predator size. Science 308:1934–1937. https://doi.org/10.1126/science.1108841

Thorogood R, Davies NB (2012) Cuckoos combat socially transmitted defenses of Reed Warbler hosts with a plumage polymorphism. Science 337:578–580. https://doi.org/10.1126/science.1220759

Thorogood R, Davies NB (2013) Hawk mimicry and the evolution of polymorphic cuckoos. Chin Birds 4:39–50. https://doi.org/10.5122/cbirds.2013.0002

Tilgar V, Saag P, Kulavee R, Mand R (2010) Behavioral and physiological responses of nestling pied flycatchers to acoustic stress. Horm Behav 57:481–487. https://doi.org/10.1016/j.yhbeh.2010.02.006

Tillmann JE (2009) An ethological perspective on defecation as an integral part of anti-predatory behaviour in the grey partridge (*Perdix perdix* L.) at night. J Ethol 27:117–124. https://doi.org/10.1007/s10164-008-0094-z

Tinbergen N (1948) Social releasers and the experimental method required for their study. Wilson Bull 60:6–51

Tinbergen N (1951) The study of instinct. Clarendon Press, New York

Trail PW (1987) Predation and antipredator behaviour at Guianan Cock-of-the-rock leks. The Auk 104:496–507. https://doi.org/10.2307/4087549

Trnka A, Grim T (2013) Color plumage polymorphism and predator mimicry in brood parasites. Front Zool 10:1–10. https://doi.org/10.1186/1742-9994-10-25

Trnka A, Prokop P (2012) The effectiveness of hawk mimicry in protecting cuckoos from aggressive hosts. Anim Behav 83:263–268. https://doi.org/10.1016/j.anbehav.2011.10.036

Trnka A, Prokop P, Grim T (2012) Uncovering dangerous cheats: how do avian hosts recognize adult brood parasites. PLoS One 7:e37445. https://doi.org/10.1371/journal.pone.0037445

Trnka A, Trnka M, Grim T (2015) Do rufous common cuckoo females indeed mimic a predator? An experimental test. Biol J Linn Soc 116:134–143. https://doi.org/10.1111/bij.12570

Troje NF, Huber L, Loidolt M, Aust U, Fieder M (1999) Categorical learning in pigeons: the role of texture and shape in complex static stimuli. Vision Res 39:353–366. https://doi.org/10.1016/S0042-6989(98)00153-9

Tryjanowski P, Goławski A (2004) Sex differences in nest defence by the red-backed shrike *Lanius collurio*: effects of offspring age, brood size, and stage of breeding season. J Ethol 22:13–16. https://doi.org/10.1007/s10164-003-0096-9

Tvardíková K, Fuchs R (2010) Tits use a modal completion in predator recognition: a field experiment. Anim Cogn 13:609–615. https://doi.org/10.1007/s10071-010-0311-3

Tvardíková K, Fuchs R (2011) Do birds behave according to dynamic risk assessment theory? A feeder experiment. Behav Ecol Sociobiol 65:727–733. https://doi.org/10.1007/s00265-010-1075-0

Tvardíková K, Fuchs R (2012) Tits recognize the potential dangers of predators and harmless birds in feeder experiments. J Ethol 30:157–165. https://doi.org/10.1007/s10164-011-0310-0

Ullman S, Vidal-Naquet M, Sali E (2002) Visual features of intermediate complexity and their use in classification. Nat Neurosci 5:682–687. https://doi.org/10.1038/nn870

van den Hout PJ, Piersma T, Dekinga A, Lubbe SK, Visser GH (2006) Ruddy turnstones Arenaria interpres rapidly build pectoral muscle after raptor scares. J Avian Biol 37:425–430. https://doi.org/10.1111/j.0908-8857.2006.03887.x

Van den Hout PL, Mathot KJ, Maas LRM, Piersma T (2010a) Predator escape tactics in birds: linking ecology and aerodynamics. Behav Ecol 21:16–25. https://doi.org/10.1093/beheco/arp146

Van den Hout AJM, Eens M, Darras VM, Pinxten R (2010b) Acute stress induces a rapid increase of testosterone in a songbird: implications for plasma testosterone sampling. Gen Comp Endocr 168:505–510. https://doi.org/10.1016/j.ygcen.2010.06.012

van der Veen IT (1999) Effects of predation risk on diurnal mass dynamics and foraging routines of yellowhammers (*Emberiza citronella*). Behav Ecol 10:545–551. https://doi.org/10.1093/beheco/10.5.545

Van Dongen S, Lens L, Matthysen E (2001) Developmental instability in relation to stress and fitness in birds and moths studied by the Laboratory of Animal Ecology of the University of Antwerp. Belg J Zool 131:59–64

Van Hamme LJ, Wasserman EA, Biederman I (1992) Discrimination of contour-deleted images by pigeons. J Exp Psychol Anim Behav Process 18:387–399. https://doi.org/10.1037/0097-7403.18.4.387

Vaughan W, Greene SL (1984) Pigeon visual memory capacity. J Exp Psychol Anim Learn Cogn 10:256–271. https://doi.org/10.1037/0097-7403.10.2.256

Vaughan W, Herrnstein RJ (1987) Choosing among natural stimuli. J Exp Anal Behav 47:5–16. https://doi.org/10.1901/jeab.1987.47-5

Veen T, Richardson DS, Blaakmeer K, Komdeur J (2000) Experimental evidence for innate predator recognition in the Seychelles warbler. Proc Biol Sci 267:2253–2258. https://doi.org/10.1098/rspb.2000.1276

Veselý P, Buršíková M, Fuchs R (2016) Birds at the winter feeder do not recognize an artificially coloured predator. Ethology 122:937–944. https://doi.org/10.1111/eth.12565

Von Fersen L, Lea SEG (1990) Category discrimination by pigeons using five polymorphous features. J Exp Anal Behav 54:69–84. https://doi.org/10.1901/jeab.1990.54-69

Walters JR (1990) Anti-predatory behavior of lapwings: field evidence of discriminative abilities. Wilson Bull 102:49–70

Wasserman EA, Kirkpatrick-Steger K, Van Hamme LJ, Biederman I (1993) Pigeons are sensitive to the spatial organization of complex visual stimuli. Psychol Sci 4:336–341. https://doi.org/10.1111/j.1467-9280.1993.tb00575.x

Watanabe S (2001a) Discrimination of cartoons and photographs in pigeons: effects of scrambling of elements. Behav Process 53:3–9. https://doi.org/10.1016/S0376-6357(00)00139-X

Watanabe S (2001b) Van Gogh, Chagall and pigeons: picture discrimination in pigeons and humans. Anim Cogn 4:147–151. https://doi.org/10.1007/s100710100112

Watanabe S, Sakamoto J, Wakita M (1995) Pigeons' discrimination of paintings by Monet and Picasso. J Exp Anal Behav 63:165–174. https://doi.org/10.1901/jeab.1995.63-165

Watson JB (1913) Psychology as the behaviorist views it. Psychol Rev 20:158–177. https://doi.org/10.1037/h0074428

Watve M, Thakar J, Kale A, Puntambekar S, Shaikh I, Vaze K, Jog M, Paranjape S (2002) Bee-eaters (Merops orientalis) respond to what a predator can see. Anim Cogn 5:253–259. https://doi.org/10.1007/s10071-002-0155-6

Weatherhead PJ (1989) Nest defence by song sparrows: methodological and life history considerations. Behav Ecol Sociobiol 25:129–136. https://doi.org/10.1007/BF00302929

Weidinger K (2002) Interactive effects of concealment, parental behaviour and predators on the survival of open passerine nests. J Anim Ecol 71:424–437. https://doi.org/10.1046/j.1365-2656.2002.00611.x

Weidinger K (2010) Foraging behaviour of nest predators at open-cup nests of woodland passerines. J Ornithol 151:729–735. https://doi.org/10.1007/s10336-010-0512-1

Welbergen JA, Davies NB (2008) Reed warblers discriminate cuckoos from sparrowhawks with graded alarm signals that attract mates and neighbours. Anim Behav 76:811–822. https://doi.org/10.1016/j.anbehav.2008.03.020

Welbergen JA, Davies NB (2009) Strategic variation in mobbing as a front line of defense against brood parasitism. Curr Biol 19:235–240. https://doi.org/10.1016/j.cub.2008.12.041

Welbergen JA, Davies NB (2011) A parasite in wolf's clothing: hawk mimicry reduces mobbing of cuckoos by hosts. Behav Ecol 22:574–579. https://doi.org/10.1093/beheco/arr008

Wheatcroft D, Gallego-Abenza M, Qvarnström A (2016) Species replacement reduces community participation in avian antipredator groups. Behav Ecol 27:1499–1506. https://doi.org/10.1093/beheco/arw074

Whittingham ML, Butler SJ, Quinn JL, Cresswell W (2004) The effect of limited visibility on vigilance behaviour and speed of predator detection: implications for the conservation of granivorous passerines. Oikos 106:377–385. https://doi.org/10.1111/j.0030-1299.2004.13132.x

Wiklund CG (1990) The adaptive significance of nest defence by merlin, Falco columbarius, males. Anim Behav 40:244–253. https://doi.org/10.1016/S0003-3472(05)80919-4

Winkler DW (1992) Causes and consequences of variation in parental defence behaviour by tree swallows (Tachycineta bicolor). Condor 94:502–520. https://doi.org/10.2307/1369222

Winkler DW (1994) Anti-predator defence by neighbours as a responsive amplifier of parental defence in tree swallows. Anim Behav 47:595–605. https://doi.org/10.1006/anbe.1994.1083

Wong S (1999) Visual predator recognition in Australian brush-turkey (*Alectura lathami*) hatchlings. In: Dekker RWRJ, Jones DN, Benshemesh J (eds) Proceedings of the third international megapode symposium, Nhill, Australia, December 1997. Zoologische Verhandelingen, Leiden

Yang C, Wang L, Cheng S-J, Hsu Y-C, Liang W, Møller AP (2014) Nest defenses and egg recognition of yellow-bellied prinia against cuckoo parasitism. Naturwissenschaften 101:727–734. https://doi.org/10.1007/s00114-014-1209-8

Ylönen H, Eccard JA, Jokinen I, Sundell J (2006) Is the antipredatory response in behaviour reflected in stress measured in faecal corticosteroids in a small rodent? Behav Ecol Sociobiol 60:350–358. https://doi.org/10.1007/s00265-006-0171-7

Yorzinski JL, Vehrencamp SL (2009) The effect of predator type and danger level on the mob calls of the American crow. Condor 111:159–168. https://doi.org/10.1525/cond.2009.080057

Young ME, Peissig JJ, Wasserman EA, Biederman I (2001) Discrimination of geons by pigeons: the effects of variations in surface depiction. Anim Learn Behav 29:97–106. https://doi.org/10.3758/BF03192819

Yu JP, Wang LW, Xing XY, Yang CC, Ma JH, Moller AP, Wang HT, Liang W (2016) Barn swallows (*Hirundo rustica*) differentiate between common cuckoo and sparrowhawk in China: alarm calls convey information on threat. Behav Ecol Sociobiol 70:171–178. https://doi.org/10.1007/s00265-015-2036-4

Yu J, Xing X, Jiang Y, Liang W, Wang H, Møller AP (2017) Alarm call-based discrimination between common cuckoo and Eurasian sparrowhawk in a Chinese population of great tits. Ethology 123:542–550. https://doi.org/10.1111/eth.12624

Zaccaroni M, Ciuffreda M, Paganin M, Beani L (2007) Does an early aversive experience to humans modify antipredator behaviour in adult Rock partridges. Ethol Ecol Evol 19:193–200. https://doi.org/10.1080/08927014.2007.9522561

Zorina ZA, Obozova TA (2011) New data on the brain and cognitive abilities of birds. Zool Zh 90:784–802. https://doi.org/10.3758/BF03192819

Printed in the United States
By Bookmasters